AF575447

SAP PRESS is a joint initiative of SAP and Rheinwerk Publishing. The know-how offered by SAP specialists combined with the expertise of Rheinwerk Publishing offers the reader expert books in the field. SAP PRESS features first-hand information and expert advice, and provides useful skills for professional decision-making.

SAP PRESS offers a variety of books on technical and business-related topics for the SAP user. For further information, please visit our website: *www.sap-press.com*.

Abassin Sidiq
SAP Analytics Cloud (3rd Edition)
2024, 421 pages, hardcover and e-book
www.sap-press.com/5753

Bertram, Charlton, Hollender, Holzapfel, Licht, Paduraru
Designing Dashboard with SAP Analytics Cloud
2021, 344 pages, hardcover and e-book
www.sap-press.com/5235

Antoine Chabert, David Serre
SAP Analytics Cloud: Predictive Analytics
2024, 315 pages, hardcover and e-book
www.sap-press.com/5771

Das, Berner, Shahani, Harish
SAP Analytics Cloud: Financial Planning and Analysis
2022, 468 pages, hardcover and e-book
www.sap-press.com/5486

Banda, Chandra, Gooi
SAP Business Technology Platform (2nd Edition)
2024, 729 pages, hardcover and e-book
www.sap-press.com/5919

Josef Hampp, Jan Lang

Application Development with SAP® Analytics Cloud

Editor Rachel Gibson
Acquisitions Editor Hareem Shafi
German Edition Editor Nicole Hohmann
Copyeditor Doug McNair
Cover Design Graham Geary
Photo Credit iStockphoto: 171573065/© real444
Layout Design Vera Brauner
Production Eric Wyche
Typesetting III-Satz, Germany
Printed and bound in the United States of America, on paper from sustainable sources

ISBN 978-1-4932-2640-5

1st edition 2025

Library of Congress Cataloging-in-Publication Control Number: 2024041619

Contents at a Glance

Contents

Introduction

Data-driven decisions are becoming increasingly relevant in modern companies. In addition to the right data storage and provision, the presentation of information is crucial for success. Traditional dashboards no longer cover the basic needs of some end users, and that is why analytical applications are becoming increasingly important to companies.

This is a practical manual for getting started on story design with scripting in SAP Analytics Cloud. We introduce the *advanced mode*, which is the development environment for scripting capabilities in SAP Analytics Cloud. We also show you how to use the advanced mode to develop your own stories with custom logic and widgets, and we'll teach you'll how to make these stories available to users in your company and monitor the stories' use and performance.

Target group

The target groups for this book include the following:

- Business intelligence (BI) experts
- Consultants
- Developers

Structure of This Book

In the first two chapters, we introduce the fundamental topics of analytical applications, SAP Analytics Cloud's development environment, and the Cloud's importance within the SAP strategy. Then, we demonstrate practical application development, using examples. We present the standard widgets and application programming interfaces (APIs) available for BI and planning applications, and we explain their use with short scripting examples.

We then use development examples to show you how to build your own story with scripting capabilities. We make a distinction between BI stories and planning stories, and we show you how to integrate predictive analytics functions into your own stories.

Here's a chapter-by-chapter summary of the topics we cover:

In Chapter 1, "Introduction to Scripting," we introduce the topic of analytical applications, and we position this topic within the SAP strategy.

In Chapter 2, "Basics of SAP Analytics Cloud," we explain the basic functionality of SAP Analytics Cloud and describe the advanced mode.

In Chapter 3, "Development of Business Intelligence Applications," we explain standard widgets and APIs in the context of BI stories, and we use programming examples to illustrate them. We also highlight the predictive functionalities of SAP Analytics Cloud.

In Chapter 4, "Development of Planning Applications," we explain standard widgets and APIs in the context of planning stories. In addition to short programming examples, we look at a larger planning application.

In Chapter 5, "Custom Widgets," we teach you how to develop and integrate your own widgets. This topic is particularly interesting for advanced developers.

In Chapter 6, "Provision and Operation of Stories with Scripting," we deal with the provision and operation of stories. To this end, SAP Analytics Cloud provides functions for publishing and monitoring stories.

Finally, in Chapter 7, "Outlook," we look at the SAP Analytics Cloud roadmap and show you what our vision of the future looks like.

Actuality

We would also like to take this opportunity to point out that we finished writing this book in the summer of 2024. Due to the large number of updates that regularly appear in SAP Analytics Cloud, we adapted and updated this book several times during the writing process—but unfortunately, we can't guarantee that all functions, descriptions, and screenshots will still be up to date when you read this book.

Download material

So that you can follow the examples in this book and try them out yourself immediately, you can download the programming example from *www.sap-press.com/5944* in the **Product supplements** section.

Box symbols

We use the following symbols in the text to draw your attention to important information and make it easier for you to work with this book:

Tip

This symbol indicates tips and hints from professional practice that provide practical recommendations that can make your work easier.

[!]

Warning

This symbol draws your attention to special features that you should be aware of. It also warns you of common mistakes or problems that may occur.

Note

This symbol appears in boxes where you'll find information on further topics or important content that you should remember.

Acknowledgments

Finally, we would like to thank all the people without whom this book would not exist.

We would like to thank the team at Rheinwerk Verlag (our editors Janina Schweitzer and Nicole Hohmann) for their confidence in us during work on the first edition of this book and for accompanying us along the way. We would also like to thank the US team, especially Rachel Gibson and Hareem Shafi, for their vision in translating and updating the book for the English-language market. Their support and cooperation were crucial in making this project a reality, so thanks to them for their trust and dedication to this endeavor. We would also like to thank our readers for their positive feedback and comments on this book.

You are also welcome to follow and contact us on our LinkedIn profiles, which we share in Figure 1 and Figure 2. We regularly share news there.

We hope you enjoy reading and learning from this book!

Figure 1 LinkedIn Profile of Jan Lang

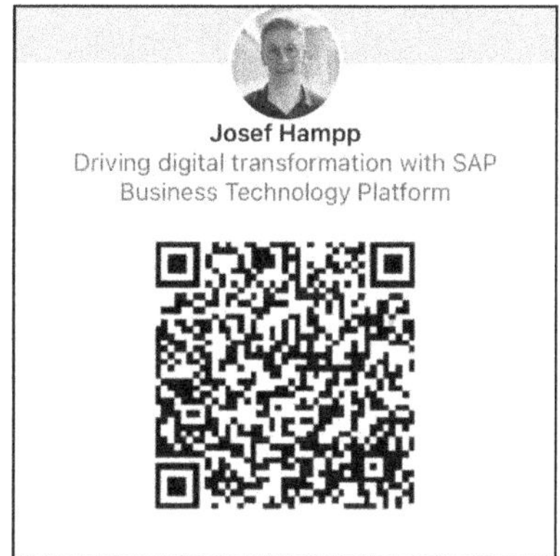

Figure 2 LinkedIn Profile of Josef Hampp

Chapter 1
Introduction to Scripting

This chapter explains the basics of analytics. You'll learn about the basic concepts and the importance of scripting in analytics, and toward the end of this chapter, we'll consider SAP's strategy and how SAP positions itself in this area.

Analytical applications are important components in the decision-making process in companies. They allow users to obtain fact-based answers to important business questions.

In Section 1.1, "What Is Analytics?", we define the term *analytical application* and position it within the overarching topic of analytics. In Section 1.2, "Analytical Software," we explain specific requirements for analytical software and when you should use it, and we describe typical characteristics of an analytical application. In Section 1.3, "Positioning Dashboards with Scripting within SAP's Analytics Strategy," we look at SAP's analytics strategy, and we introduce SAP Analytics Cloud, which is SAP's strategic analytics solution.

1.1 What Is Analytics?

Definition

There are several definitions and interpretations of the terms *analytics* and *analytical software*, so we'll first explain how we understand them.

Business applications

Figure 1.1 shows a simplified architecture of an analytical IT system. This consists of three layers, and the lowest layer consists of *enterprise applications* that a company uses. Examples of enterprise applications are business applications like *enterprise resource planning* (ERP) applications, specific applications for a business area (e.g., human resources), and applications that help with customer interactions in *customer relationship management* (CRM). Business applications process all the data required to control a company. To gain end-to-end insight, an organization needs to relate this application data. This requires the implementation of an additional data management application called a data warehouse.

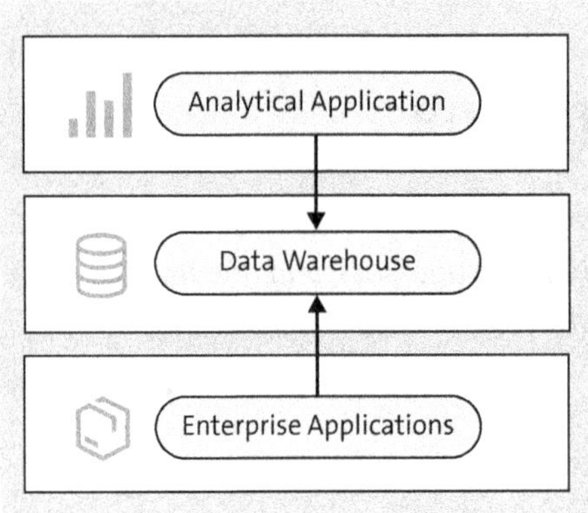

Figure 1.1 Simplified Architecture of Analytical IT System

Data warehouse

The middle layer of the architecture consists of the *data warehouse*, which handles a diverse range of tasks that include data integration, data preparation, the creation of semantic data models, and the historization of enterprise data. From the perspective of analytical software, the most important task of a data warehouse is to provide semantic data models that users can consume immediately.

Analytical software

The top layer of the architecture is the *analytical application*, with which users access the data in the data warehouse. Over time, people in the field have developed different terms for the capabilities of analytical applications, and some of these terms complement each other or overlap in terms of what they cover. Some of these terms are as follows:

- Descriptive analytics
- Predictive analytics
- Augmented analytics

Descriptive analytics

The term *descriptive analytics* describes the evaluation and presentation of company data, which often involves presenting data not in its pure form (e.g., not in table form but graphically). Various types of visualization are available for this purpose, depending on the type of information to be communicated. Descriptive analyses usually involve company data generated during daily business.

Predictive analytics

The term *predictive analytics* refers to the use of available data and the creation of predictive models based on it. *Machine learning* (ML) is a type of predictive analytics that involves the automated processing of data by algorithms, and it is used to create forecasts based on actual data. The results of such forecasts can then be integrated back into a dashboard.

Augmented analytics

The term *augmented analytics* refers to various capabilities of a system that are mainly intended to facilitate the handling of the system. These include the automated creation of dashboards or predictive models as well as their interactions with the system. *Natural language processing* (NLP) lets users use many applications in natural language.

Often, companies omit the distinctions among the above terms and only speak of analytical systems or analytical applications. This is for simplicity's sake and also because current software solutions cover all of the aforementioned capabilities.

Business intelligence

Business intelligence (BI) is another common term. Organizations use BI applications to make data-driven decisions. In their early days, BI applications were mostly management applications, but the increasing availability of data, data democratization, and new technologies has led to a widening of the BI user base. In modern companies, BI applications are usually available to all employees, and as a result, BI has lost its management focus. Nowadays, the term *BI* is often used synonymously with the terms *analytics* and *analytical software*.

Planning applications

Enterprise planning is the third core capability of analytical systems (after visualizing enterprise data and calculating forecasts). You can use *planning applications* to implement decisions made with BI dashboards. Planning applications are essential for enterprise management, and the fields where you can use these applications include financial planning, personnel planning, and production planning. For more information about developing planning applications with SAP Analytics Cloud, see Chapter 4.

1.2 Analytical Software

Different analytical software user groups have different requirements for how to present analysis results and predictions. For this reason, over the years, user groups have established various forms of representation that complement each other, which is why companies usually don't exclude any form of representation form from their analytics landscape. The ways in which insights and data are presented range from strongly number-based representations, such as tables, to visualizations that represent a measure with multiple dimensions. We'll briefly introduce you to some frequently used display formats in the following sections.

1.2.1 Reports

Reports

Analytical software offers you many options for presenting data and analysis. In the past, for example, companies have often had a strong reporting culture. A *report* is usually a static document that is made available to a user as a data consumer. The person who consumes reports is usually not responsible for creating them, and reports can be made available or shared directly using analytical software (e.g., as PDF files). Since reports are often exported, they are often presented in a highly structured way for ease of use and comprehension. If a report is exported as a PDF file or in print form, it can (for example) be created in DIN A4 format, which facilitates readability in physical form.

Figure 1.2 shows the first page of a report for a company operating in the North American market. Such reports are produced with a high level of detail so that experienced readers can analyze and understand the situation the report presents at a glance. Tables offer the advantage of letting users easily break down numbers across different dimensions (time, products, locations, and so on) or inserting new calculations (totals, differences, percentage changes, and so on). The compact representation is very advantageous to users here.

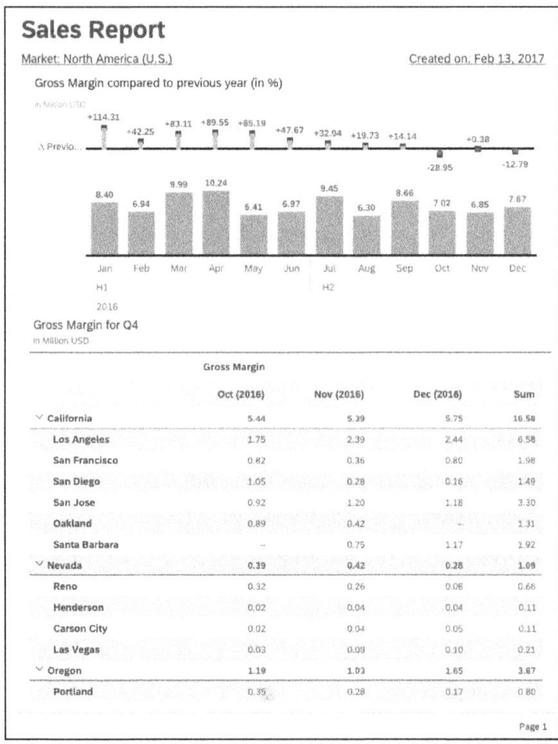

Sales Report

Market: North America (U.S.) Created on: Feb 13, 2017

Gross Margin compared to previous year (in %)

Gross Margin for Q4

in Million USD

	Gross Margin			
	Oct (2016)	Nov (2016)	Dec (2016)	Sum
California	5.44	5.39	5.75	16.58
Los Angeles	1.75	2.39	2.44	6.58
San Francisco	0.82	0.36	0.80	1.98
San Diego	1.05	0.28	0.16	1.49
San Jose	0.92	1.20	1.18	3.30
Oakland	0.89	0.42	-	1.31
Santa Barbara	-	0.75	1.17	1.92
Nevada	0.39	0.42	0.28	1.09
Reno	0.32	0.26	0.08	0.66
Henderson	0.02	0.04	0.04	0.11
Carson City	0.02	0.04	0.05	0.11
Las Vegas	0.03	0.09	0.10	0.21
Oregon	1.19	1.03	1.65	3.87
Portland	0.35	0.28	0.17	0.80

Page 1

Figure 1.2 Example of Financial Statement

1.2.2 Dashboards

Dashboard

Thanks to technological advances since the turn of the millennium, interactive *dashboards* have become increasingly important. Within an analysis program, you can call various reports and manipulate the displayed data volume with various interactions. Examples of such interactions are as follows:

- Drill-down or drill-up using dimension hierarchies
- Filter setting
- Geospatial navigation
- Operation of control elements

You can use such interactions to make dashboards simpler. You also don't have to present all the information—instead, you can display selected data in aggregated form. Figure 1.3 shows a simple dashboard that displays relevant data clearly in various formats. An advantage of dashboards is that their format (i.e., the display as a pie chart or in tiles) is dynamically aligned with the content or behaves responsively to the display screen. At the same time, you can adjust the format to meet your individual requirements.

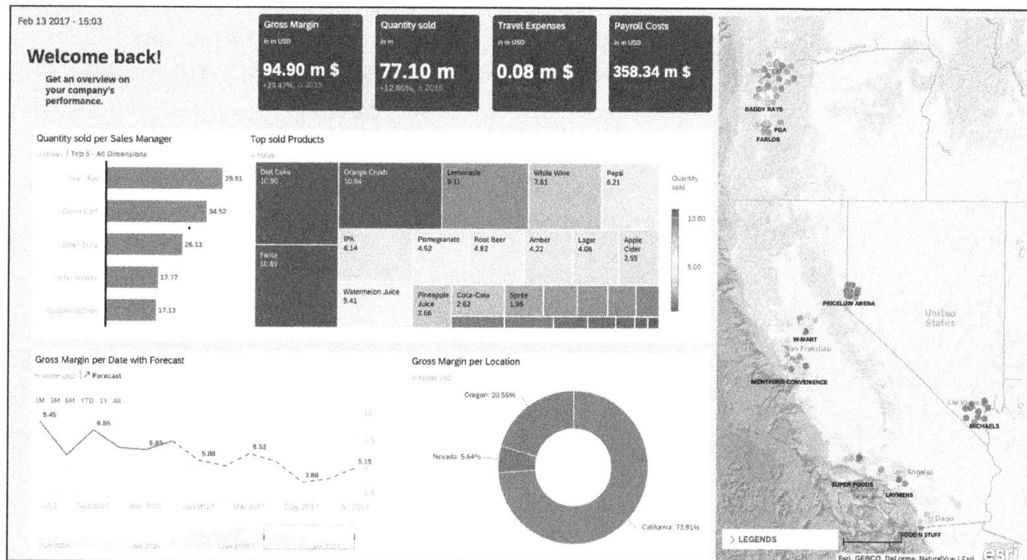

Figure 1.3 Example of Dashboard

The flexibility of a dashboard's format is enhanced by the fact that users usually consume dashboards via an analysis program and not via exports. Users can install analysis programs locally, on their computers, or access them on the internet. The latter is the more modern way to access such programs, via *software-as-a-service* (SaaS).

When creating dashboards, we make a distinction between two paradigms:

- Self-service dashboards
- Dashboards with scripting

1.2.3 Self-Service Dashboards

Self-service dashboards

The self-service dashboard paradigm holds that users should be able to create their own visualizations and dashboards without the help of third parties. No previous IT knowledge (for example, the mastery of a specific programming language) should be required, users should be able to create visualizations in a supported environment, and the analysis program should provide an editor with objects that can be created and changed. In addition, the system should issue supporting notes and warnings during creation, and it should also support design principles that simplify and explain how to use the program. Dashboard creators often follow the *what you see is what you get* (WYSIWYG) principle here, so that users can directly see how their adjustments affect the dashboard (e.g., when moving or resizing objects using drag-and-drop mechanisms).

Self-service environments are important tools for businesses. They allow employees to work independently with data to find answers to their business questions, and this often saves everyone involved time. However, there are also restrictions. Thanks to the guided design time and the what you see is what you get principle, users can map simple interaction logics, but when implementing more complex interactions or business logics, users may reach limits. Dashboards with scripting are used for the technical implementation of scenarios where users reach limits.

1.2.4 Dashboards with Scripting

Dashboards with scripting are dashboards that go beyond the scope of self-service dashboards. You can build on self-service dashboard creation by supplementing it with enhanced functions, such as executing script elements.

Necessary capabilities

Unlike self-service dashboards, which can be created by any employee in a company, dashboards with scripting are often built by dedicated experts who create different applications for the company. To develop such applications, experts use prior knowledge or skills (e.g., the ability to write program code) that not all employees have or require in their daily tasks.

Due to their high degree of customization, dashboards with scripting offer a wide range of applications, and they can help users examine and visualize data in a wide variety of ways. Users can create table-centric representations (as in a report) or data visualizations (as in a self-service dashboard). They can also work with data in dashboards with scripting by triggering or processing workflows and implementing planning applications.

1.3 Positioning Dashboards with Scripting within SAP's Analytics Strategy

Analytics is important to SAP as a leading provider of enterprise software, so in this section, we'll explain SAP's analytics strategy and give you an introduction to SAP Analytics Cloud.

1.3.1 SAP's Analytics Strategy

SAP defines *analytics* as the intelligent use of data—that is, the interaction of data visualization, BI, business planning, and augmented analytics. In the past in this area, SAP has focused primarily on the on-premise solution *SAP BusinessObjects*, which isn't just a solution but a platform that combines various applications for different application areas.

Statement of direction

With its development of SAP Analytics Cloud, however, SAP reconsidered this approach. *SAP Analytics Cloud* is an integrated analytics solution that can be licensed as SaaS. We refer to SAP Analytics Cloud as *integrated* because it can be used to map holistic processes using planning, prediction, and visualization. SAP clearly positions SAP Analytics Cloud as the strategic analytics solution in its statement of direction. This means that a large number of future innovations will be developed and released exclusively for SAP Analytics Cloud, and it also means that SAP Analytics Cloud will be the analytical user interface of SAP solutions. SAP Analytics Cloud is embedded in all strategic SAP products as a reporting interface for operational reporting.

1.3.2 SAP Analytics Cloud

In this section, we'll show you the key characteristics of SAP Analytics Cloud, which we divide into three levels:

- Top level: Access options for users
- Middle level: Core capabilities
- Lower level: Data integration interfaces

We'll discuss each level in the following sections.

Access Options for Users

The top level involves the technical ability of users to access SAP Analytics Cloud. Various access options are available. As this is a cloud application, you can access SAP Analytics Cloud directly using a dedicated URL, but we don't recommend that you use that option on mobile devices. Instead, we recommend that you use SAP's native apps that are optimized for use on mobile devices and are available free of charge for iOS and Android devices in their respective app stores.

The top level is complemented by the *catalog*, which is a tool for organizing all analytical artefacts of a company. In the catalog, you can store all visualization artefacts and data objects of SAP Analytics Cloud as well as uploaded files or links to external content. As a central portal, the catalog is the entry point for users. You can maintain the catalog yourself, or content will be made available to you. The catalog elements are displayed as cards, and individual users and teams can use individual display and read authorizations for each card.

Microsoft Office add-ins

In addition, you have the option of integrating SAP Analytics Cloud with Microsoft Office applications via add-ins. You can use add-ins to, for example, import data as a table into Microsoft Excel and process it further there or add visualizations from a story to a slide in Microsoft PowerPoint. You can install these add-ins in the relevant Microsoft Office applications and add the corresponding functions to the applications.

Core Capabilities

The core capabilities of SAP Analytics Cloud are positioned as key elements in Figure 1.4. These are divided into two areas of competence:

- Analytics and BI
- Enterprise planning

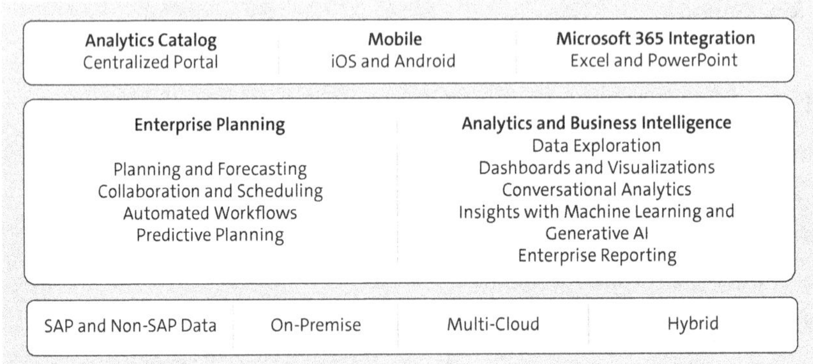

Figure 1.4 SAP Analytics Cloud (Source: SAP)

Business intelligence

BI includes all functions from interactive data exploration to the creation of dashboards and reports with meaningful representations. In addition, this area contains the publication and provision of your own objects for other users.

Analytics

The complementary term *analytics* describes a very broad range of functions, including the interaction of the system with natural language, the automation of processes such as reporting, and the use of predictive models to create forecast calculations. More generally, analytics (a.k.a. *augmented analytics*) include all features that use intelligent algorithms to simplify the way users work with SAP Analytics Cloud.

Enterprise planning

Enterprise planning includes functions such as entering and processing plan values, allocating to planning objects, mapping planning processes, and forecasting planning data.

As you can see, these areas overlap in their individual functions and core capabilities. In both analytics and enterprise planning, forecast models are used to calculate forecast data. This is intentional because SAP wants to offer added value in holistic analytics processes through a deep integration of the functions with SAP Analytics Cloud. We separate these areas of competence here only to provide a simplified description on paper. For users, there are no boundaries between the individual areas; they are just the functions provided by SAP Analytics Cloud.

As a modern solution, SAP Analytics Cloud offers several interaction options for the various needs of end users (see Figure 1.5).

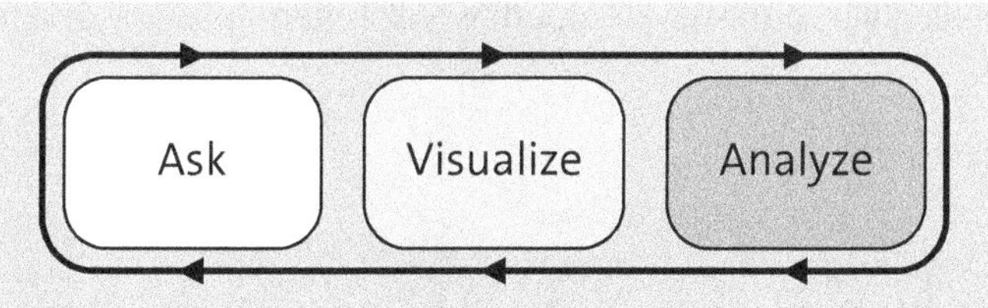

Figure 1.5 Interaction Possibilities in SAP Analytics Cloud

Just ask

Occasional users who are only interested in one or a few data points can ask SAP Analytics Cloud a question in natural language and receive a specific answer. This is the *just ask functionality*, and with it, you can quickly and easily query data points.

Story

Use cases relating to reporting, dashboarding, analytics, augmented analytics, or enterprise planning are implemented in one artifact. In SAP Analytics Cloud, an artifact is referred to as a *story*, and it can be created in self-service dashboards or extended by scripting.

Data analyzer

Stories often present data as a visualization on an aggregated level. Users, who are often experts in a specialized field, also want to view data at a granular level, so SAP Analytics Cloud offers the option of viewing data at row level, in a table, and customizing such a table on an ad-hoc basis. The *data analyzer* is used for this purpose.

Data Integration Interfaces

The lower level includes the data integration interfaces of SAP Analytics Cloud. SAP Analytics Cloud offers a large number of native connectors that offer a ready-made interface for any data source. It's important to note that SAP Analytics Cloud is a solution that can integrate both SAP and non-SAP data sources. In addition to these connectors, SAP Analytics Cloud offers generic interface technologies such as SQL and open data protocol (OData), which you can use to query data from data sources for which there's no dedicated connector. On the infrastructure side, SAP Analytics Cloud has all the options for addressing data sources from on-premise installations to cloud applications.

1.4 Summary

In this chapter, we gave a brief overview of the basics of analytics and SAP's strategy in this area. We also underlined the importance of app development for analytics purposes, and we discussed how SAP is positioning SAP Analytics Cloud as a solution in this area.

In the next chapter, we'll teach you the fundamentals of SAP Analytics Cloud and dive deeply into the development environment for story design with scripting.

Chapter 2
Basics of SAP Analytics Cloud

This chapter explains the basics of SAP Analytics Cloud and introduces the integrated development environment—which we call advanced mode—and the JavaScript language scope available for scripting there. This chapter also explains the required authorizations as a prerequisite for advanced mode.

What you can expect

Now that you've taken a closer look at the topic of analytics in the previous chapter, you'll learn how to work with SAP Analytics Cloud and what requirements you must meet to work with it.

In Section 2.1, "Architecture of the System Landscape," we'll give you an overview of a possible system landscape with SAP Analytics Cloud. In Section 2.2, "Menu Navigation and Important Functions," we'll show you how to find your way around SAP Analytics Cloud. In Section 2.3, "Data Integration," and Section 2.4, "Data Modeling," you'll learn how to integrate and model data. In Section 2.5, "History of Scripting in SAP Analytics Cloud," you'll learn about the improvements SAP has made that help you as a story designer. Then, in Section 2.6, "Difference between Stories and Advanced Mode," you'll learn about the differences between a simple story and advanced mode, and in Section 2.7, "Development Environment," you'll learn about the development environment. In Section 2.8, "Authorization for Scripting with SAP Analytics Cloud," you'll gain insights into the authorizations and requirements for scripting in advanced mode, and we'll show you how to integrate the available demo stories into your tenant and migrate old objects.

2.1 Architecture of the System Landscape

When companies decide to have their cloud solutions operated by external providers such as SAP by using the SaaS model, they give up their responsibility for operating the application's infrastructure. This is because they are outsourcing this task to an external service provider, which has far-reaching effects on the architecture of the system landscape. In this section, we explain how the system landscape is changing in this respect.

2.1.1 Three-System Landscape

The operation and maintenance of IT infrastructure used to be one of the most important tasks of IT departments. Due to various requirements, several instances of the same product were operated in parallel—in other words, it was a *multisystem landscape*. A system landscape consisting of three instances has become established, as you can see in Figure 2.1.

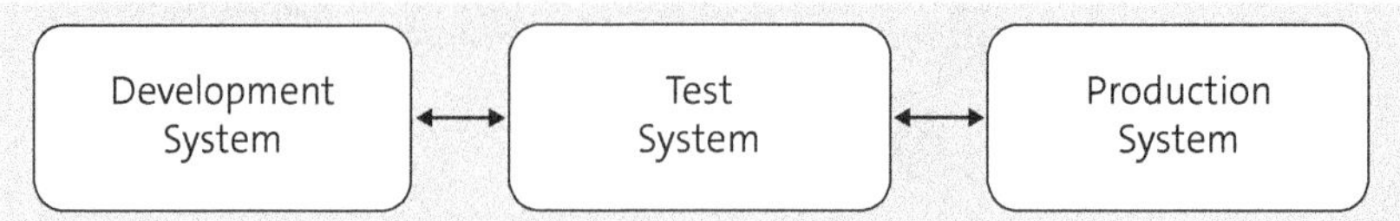

Figure 2.1 Representation of Three-System Landscape

Three-system landscape

The name of the *development system* indicates that it is an instance in which new objects can be developed. Most users in a company don't have access to this system; it is usually reserved for the relevant IT department. In the *test system*, software is tested before it's executed in the *production system*. The objects created are checked for errors in the test system. For example, in the case of on-premise installations, IT has to regularly install system updates and release changes. As long periods of time can pass between updates, they can have a major impact on the functionality of the objects. Therefore, IT first installs and checks updates on the test system before it rolls out new functions on the production system. For this reason, IT generally does not use production data in the development and test systems.

As the multisystem landscape is a universally valid paradigm, not only analytical systems but all systems managed by IT have been set up in it. Accordingly, the data warehouse and the enterprise resource planning system also exist in multiple instances, and as a result, instances from the same class are linked together. For example, the test system of the analytical system is connected to the test system of the data warehouse and can use its data.

2.1.2 Change to a Two-System Landscape

The widespread use of cloud solutions means that tasks originally performed by IT are no longer necessary. For example, IT used to be responsible for the smooth operation and availability of infrastructure and applications. However, cloud solutions are usually offered as infrastructure-as-a-service (IaaS) or SaaS solutions, and for the consumer, this means that the responsibility for operating the solution has shifted from the company to the cloud provider. This also includes installing new updates or carrying out release upgrades. As SAP Analytics Cloud is provided according to this model, the latest functions are always delivered automatically. Nevertheless, SAP defines fixed times at which the system receives updates. Two variants are available here: a biweekly and a quarterly update cycle. The quarterly update cycle is recommended by default, which means that you'll receive new innovations once a quarter.

Two-system landscape

In terms of the requirements for the system landscape, you or your IT department no longer have to perform tasks such as supporting release upgrades. Due to the reduced workload, a *two-system landscape* is recommended. The production system remains in place, and an additional instance is added for development and testing purposes (see Figure 2.2).

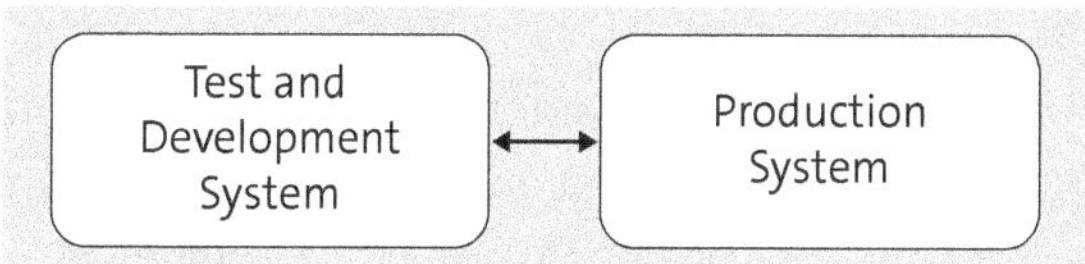

Figure 2.2 Representation of Two-System Landscape

2.2 Menu Navigation and Important Functions

After you log in to SAP Analytics Cloud, the start screen opens (see Figure 2.3).

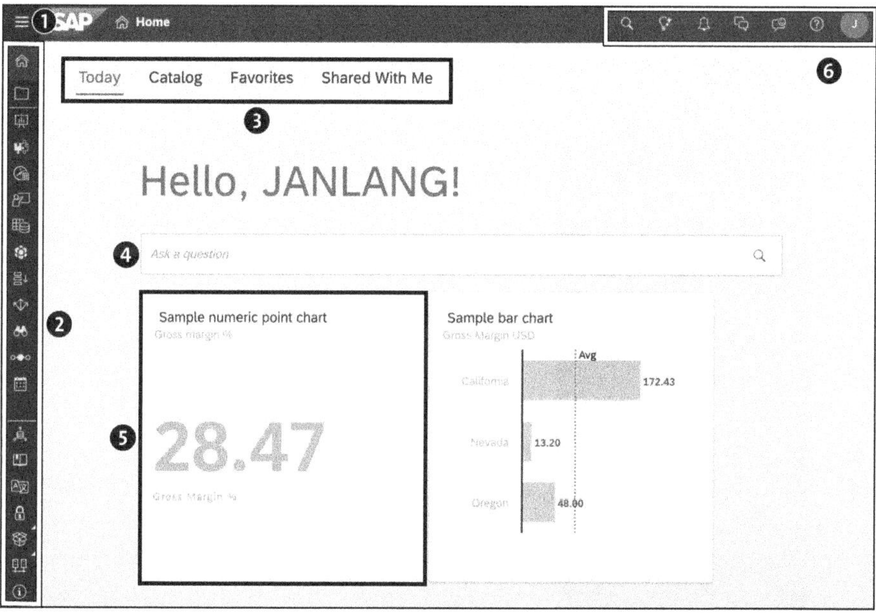

Figure 2.3 Homepage of SAP Analytics Cloud

It is divided into three general areas

- **Shell bar**
  The upper horizontal area is called the *shell bar*. It enables direct navigation to the important functions of SAP Analytics Cloud.
- **Navigation bar**
  The vertical area on the left-hand side of the screen is the *navigation bar*. This is the main menu from which you can jump to the various areas of SAP Analytics Cloud.
- **Workspace**
  The rest of the screen is the *workspace*. This is where the various functional areas of SAP Analytics Cloud are displayed and where you as a user can work.

Sidebar

Now, let's take a closer look at the individual elements of the home screen. The sidebar (Figure 2.3 ❶) is normally collapsed, and the individual menu items are only displayed as icons. If you click on the ☰ icon, the navigation bar opens (see Figure 2.4).

Figure 2.4 Expanded Navigation Bar

The navigation bar (Figure 2.3 ❷) consists of three areas:

Navigation bar

- The first area is aimed at all users. It takes you back to the start page and the **Files** page.
- The second area, **Apps**, differs depending on the user. The navigation options here result from the authorizations and roles assigned to the user.
- The third area, at the bottom, bundles all menu items relating to the administrative functions. These are only visible to users who have the corresponding authorizations.

[«]

Responsive Layout of the Navigation Bar

The navigation bar adapts to the vertical height of the screen, so it's possible that you'll see elements displayed there other than those in Figure 2.4. Menu items that can't be displayed due to low height are collected in the **More** submenu.

Tabs on the start page

The start page consists of several tabs that you can open via this navigation (see Figure 2.3). There are four tabs in total ❸. The **Today** tab lets you make individual adjustments, while the **Catalog** tab opens the catalog (more on this in Chapter 6). The **Favorites** tab shows an area in which your favorites are listed directly as a map display, and the **Shared with me** tab opens a similar list. Only objects that other users have shared with you are displayed here. Administrators can define which tabs are displayed by default when users log in, but users have the option of overwriting the administrators' default settings via the profile settings and selecting their own view.

You can use the search box ❹ as well as the lightbulb icon in the shell bar to open the *search to insight* function. This is a search function in which you can formulate questions in natural language and SAP Analytics Cloud gives you an answer based on your data model. Alternatively, the new just ask functionality can also be displayed here, depending on the settings of your administrator.

Customize home page

As the user, you can customize the start screen ❺, which is a personal area that only you can see. You can see which stories you opened last, get quick access to those objects, and save individual widgets as tiles there. This allows you to check your most important key figures as soon as you start SAP Analytics Cloud.

[»]

Individualization of the Start Page

Your start page is private. Any changes you make will only affect your personal start page, and you can't share your start page with other users.

The shell bar (Figure 2.3 ❻) at the top right is always available to you. The icons contained there allow you to use central functions, such as searching for objects, displaying notifications, and opening collaboration options. You also have the option of using the user icon to open your profile. You can do this to manage your profile settings, for example.

2.3 Data Integration

Variants of data integration

To be able to work effectively in SAP Analytics Cloud, you must first set up data integration. The solution offers you two options for working with data:

- Via a live data connection
- Via a data import connection

We'll discuss each of these next, along with the steps for creating a new connection.

2.3.1 Live Data Connections

Live data connections allow you to access data live (i.e., in real time). When a story is called up, a request is sent to the data source and the current data stock is displayed. This is particularly advantageous if you carry out analyses based on frequently changing data. The data replication step is no longer necessary, which is also interesting for data protection and security reasons, as you don't have to maintain the data outside the company network.

Live data connection

Live data connections are possible for both cloud and on-premise data sources. Figure 2.5 gives you an overview of which data sources you can use as live data sources. If you've configured the connection accordingly, you can create live data models based on this connection. No data is saved in these models; they are storage objects for the metadata of your data model (description of the key figures, description of the dimensions, hierarchies, etc.). This information is required to create stories.

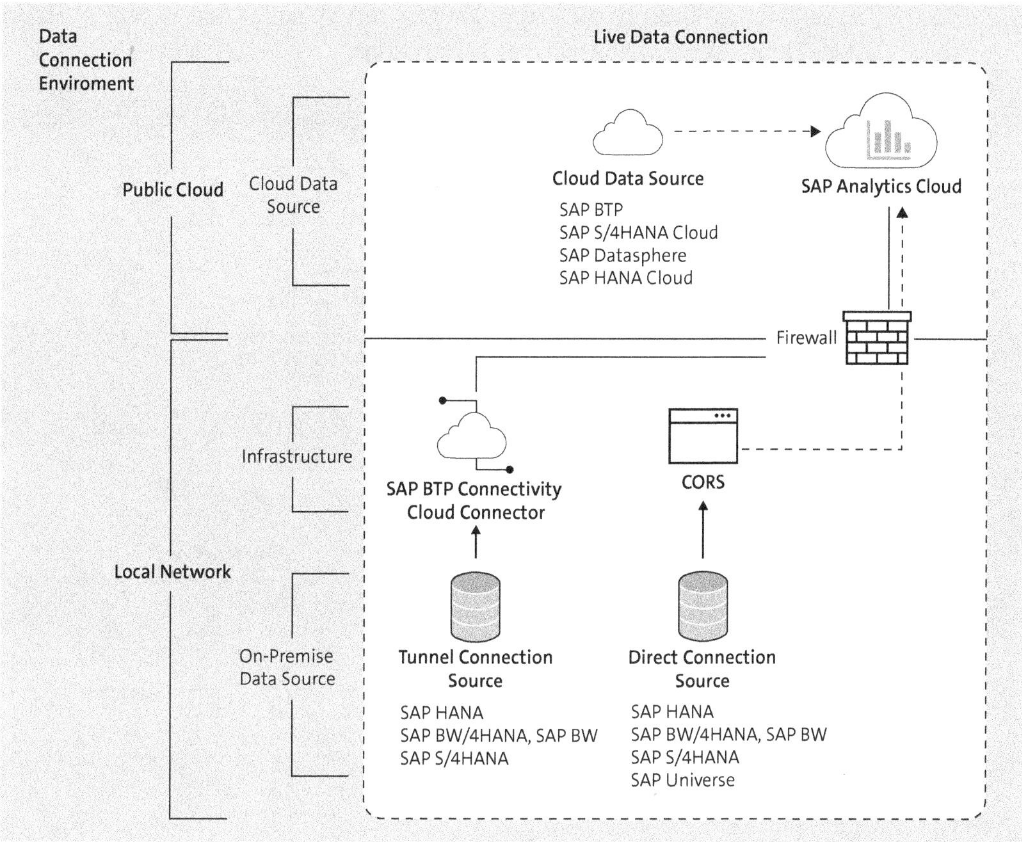

Figure 2.5 Overview Diagram for Live Data Connections

2.3.2 Data Import Connection

The *data import connection* approach is different from the live data connection approach, and the biggest difference is that in a data import connection, the data from the data source is replicated in SAP Analytics Cloud. So why go to all this effort if you could just consume the data directly via a live data connection?

Data import connection

The answer is that if you've imported data into SAP Analytics Cloud, you have access to functions that you wouldn't be able to use with a live data model. With regard to data modeling, you can enrich your model with additional semantics (e.g., enrich geo-information, add hierarchies). Furthermore, entire functional areas of SAP Analytics Cloud are only possible with import data models. This includes the entire area of corporate planning.

Another advantage of the data import connection is shown in Figure 2.6. SAP Analytics Cloud uses a variety of interfaces for data import, and in this way, significantly more data sources can be connected to SAP Analytics Cloud than with the live data connection.

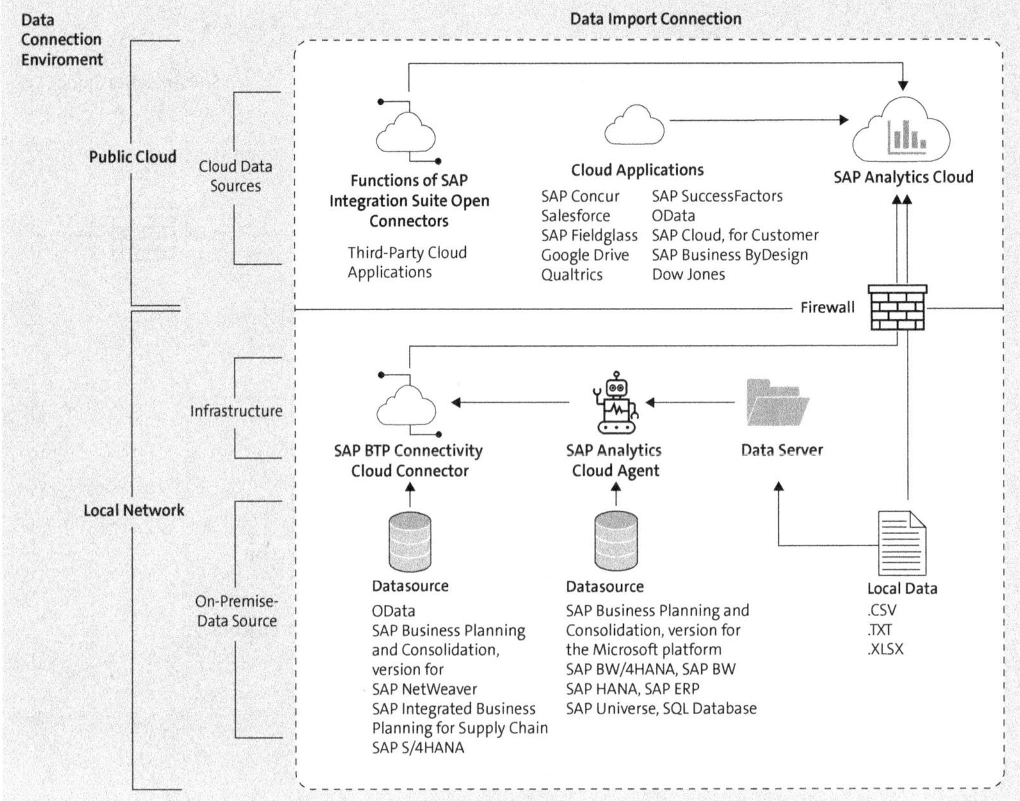

Figure 2.6 Overview Diagram for Import Connections

Data Import Connection or Live Data Connection?

Live data connection and data import connection have advantages and disadvantages, so which type of data access is right for you depends on the requirements of your analysis.

2.3.3 Creating a New Connection

Create data connection

As soon as you know which type of data connection you want to create, click the **Connections** button in the navigation bar. This opens the workspace where you can create new connections.

In this menu, you'll see a list of all existing connections, and you can select individual data import connections and share them with other users. To create a new connection, click the icon with the **+** sign. Depending on the type of connection you want to create, you may need to carry out preparatory steps.

[«]

Creating Data Connections

The steps you need to take in advance to establish a secure data connection differ depending on the data source. You can find all information, whether on premise or in a cloud data source, in the SAP Help Portal in the **Data Connections** section: *http://s-prs.co/v594401*

2.4 Data Modeling

Data management

Data modeling is an important step in preparing your data analysis. During data modeling, you can add calculations, create hierarchies, and enrich your data model with geo-information.

SAP Analytics Cloud provides you with two options for data modeling:

- Models
- Datasets

Models and datasets are two different data objects that you can use for data management. The functions available to you during data preparation and data wrangling also depend on what option you choose. We explain the two options and their strengths and weaknesses below, and whether you use a model or a dataset should always depend on your use case.

2.4.1 Datasets

Data in tables

Datasets are particularly suitable if you want to analyze data on the basis of CSV or Excel files, so they are suitable for the ad hoc analysis of data. With datasets, the data is saved in tabular form; metadata is defined separately. This structure allows you to make changes quickly in the dataset. In addition to ad hoc analyses, you can use datasets as the basis for Smart Predict.

2.4.2 Models

Models

Models are particularly suitable for handling *controlled data* (i.e., data from a source system). Models use either imported data or live data:

- **Imported data**
  If you use import models, the data is replicated to SAP Analytics Cloud. Changes to the dataset of the data source have no effect on the data in the model.

- **Live data**
  When you're using a live model, only descriptive data is stored in SAP Analytics Cloud. The actual dataset is retrieved from the source system in real time each time you use data. This means that changes to the data in the data source are also directly visible in SAP Analytics Cloud.

Advantages of the model

In contrast to data in datasets, the data in models is saved in a star schema. You can load new data into the data model via the data management of a data model created with imported data. You can either start this process manually or schedule it—in which case, the data upload happens automatically. Another advantage of the model is that you can implement row security, which allows you to define at row level which users can work with the data. This also allows you to set up more complex access authorization scenarios.

Requirements for Planning Scenarios

If you want to use planning functions with SAP Analytics Cloud, you have to work with imported data. You can't realize planning scenarios with live data.

2.5 History of Scripting in SAP Analytics Cloud

In the past, there were three types of artifacts in SAP Analytics Cloud for developing a dashboard:

- Classic stories
- Analytical applications
- Digital boardrooms

These three types of objects covered various requirements. For example, classic stories were self-service environments in which you could create meaningful dashboards in the shortest possible time. With analytical applications, you could build more complex dashboards with your own code and logic, and SAP Digital Boardroom offered you advanced navigation and display options for larger dashboards (e.g., for decision-making, for controlling a company).

SAP has now brought these objects together, and you can add a story through scripting in the new optimized design experience. The presentation mode has also been added to SAP Analytics Cloud, and it replaces SAP Digital Boardroom with its functions, without an additional license. This should result in improvements on three levels for you as a story designer:

- **Standardization**
  The introduction of the optimized story experience brings together the different development environments. When starting to create a dashboard, you no longer need to decide which development environment is best suited to which form of presentation. This is because the optimized story experience offers a uniform development environment for all application areas. Application development, self-service functionalities, and presentation development, as offered in the past with SAP Digital Boardroom, are combined in one user-friendly interface.
- **Performance and user-friendliness**
  For you, *user-friendliness* means that you can start designing in a self-service environment, and - as soon as you reach a point where you want to implement more complex logic with scripting, you can access and use the corresponding functions directly. In addition to improvements in handling the software, there are performance-related improvements that affect things like the rendering of the story, data queries to the source system, and the persistence of metadata.
- **Innovation**
  The optimized story experience is the strategic foundation for all further improvements. In future, all new functions will only be provided in optimized story experiences, so you need to convert classic stories and analytical applications to take advantage of the latest innovations.

We'll look at how to create such a story in the optimized design experience in the following sections.

2.6 Difference between Stories and Advanced Mode

Story vs. extended mode

As we already introduced in Chapter 1, Section 1.3.2, there are different display formats in SAP Analytics Cloud, and stories and their extended mode are the most important ones. Their task is to visualize information and data clearly, but there are some differences between "simple" stories and those that have been enhanced with scripts in advanced mode in SAP Analytics Cloud. The exact differences depend on whether you only use SAP Analytics Cloud as a consumer or create visualizations yourself. Depending on your perspective, you'll notice more or fewer differences.

The main difference between a story and the scripting extension is that a script contains user-defined logic that is added to the story. This option isn't available to you without advanced mode.

2.6.1 Differences for Consumers

From the perspective of the data consumer

If you only consume stories in your daily work, you should not even notice whether they have been enhanced with scripting when you work with them and evaluate your data. The widgets you use, the appearance, and the behavior are identical. In advanced mode, it's possible to store code in a story and thus provide an interactive application, and this makes it easier for you as a user to work with such stories. Only experienced users may be able to tell the difference based on these features.

2.6.2 Differences for Creators

If you're a developer working with SAP Analytics Cloud, you'll find that there are significant differences between creating a story and extending it through scripting.

Properties of a story

When you create a simple story, you probably have completely different expectations than when expanding it through scripting. Stories place a strong focus on the self-service aspect and are kept correspondingly simple; they virtually lead users by the hand when creating them. This makes it practically impossible to create a story that doesn't work.

Properties of the extension through scripting

The situation is different with scripting extensions. When you add scripts and the resulting logic, stories become fully customizable and adaptable, but this also makes them more susceptible to errors. In contrast to the creation of simple stories, the execution of scripts can fail completely or partially.

You should therefore check whether you want to add scripts to your story for your use case. On the one hand, incorporating scripts increases the maintenance effort and the risk of errors, but on the other hand, adding scripts gives you the option of creating greater usability for your end users by incorporating various logics that guide them through the story.

2.7 Development Environment

The development environment included in SAP Analytics Cloud is called *advanced mode*. We'll now look at this in detail.

Create a new story

In Section 2.2, you learned how to navigate the menu of SAP Analytics Cloud. Within this menu, there are two options for creating or opening a story:

- Via the menu
- Via the file system

You can create a new story via the **Apps** menu in the **Stories** area. Here, you have the option of creating a new story using the **Create New** button. Below this, your most recently accessed stories are displayed in the **Recent Files** area, which you can access from this view (see Figure 2.7).

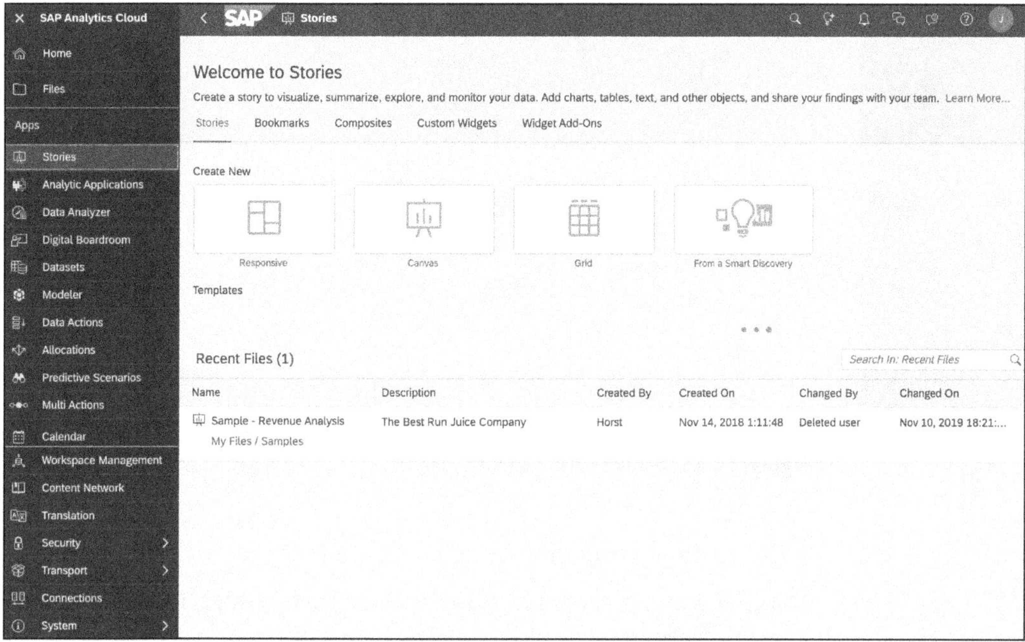

Figure 2.7 Creating Story via Menu

Switch to edit mode

If you've opened a story, it will initially be displayed in view mode. To edit the story, you must switch to edit mode in the top right-hand corner.

The second option for creating a story is creating it directly in your file system. You call up this option by selecting the **Files** area in the navigation bar (see Figure 2.8). Here, click on the **+** icon and select the **Story** option in the menu. You'll then be forwarded directly to the menu described earlier in this section, and you'll also have the option of filtering for stories in your file system.

Stories in files

You open your story via the file system directly in view mode. However, if you would like to open your story in a different mode, you can hover over the name of the story to display the three-dot menu with further actions, and you can then use this to select a different mode.

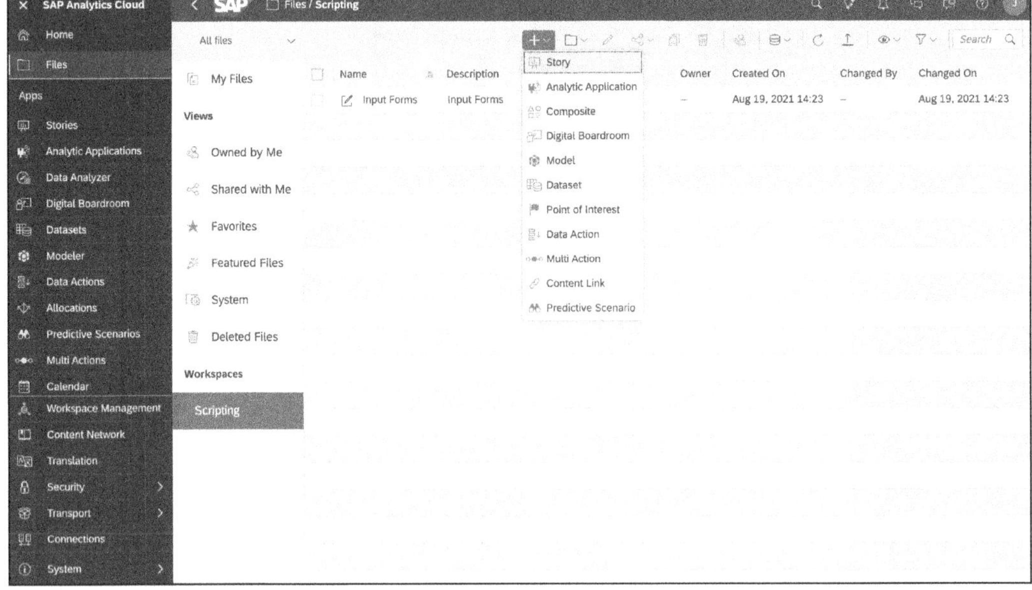

Figure 2.8 Creating Story via File System

2.7.1 General Structure

Now that you've learned how to create or open a story, let's take a closer look at advanced mode. You can activate this for your story via the puzzle icon. Simply click the icon to activate the mode for this story, and you can then expand it with scripts.

Scripting language in advanced mode

In advanced mode in SAP Analytics Cloud, you use *JavaScript* as the scripting language. The scripts are executed by the web browser in the script execution engine without prior conversion, and SAP also ensures that the scripts are validated in the development environment. Not only is a static check carried out, but the underlying model is included. This makes it possible to use input help for the dimensions and key figures in the model.

Development environment

If you create a new story or open an existing story in edit mode, you'll be sent directly to the development environment (see Figure 2.9). This is divided into different areas, which we'll look at now.

In the **File** area (Figure 2.9 ❶), you can create basic settings. These include the general story details, which you can store here. You can also define the settings for your story queries, which include batch queries and cache settings. You have extensive setting options in the subitem **View** time settings, where you can define the displayed toolbar for the different view modes of your story and also create global settings, set the navigation for your story, define the filter area, and influence the behavior of your bookmarks. You can also save and share your story here.

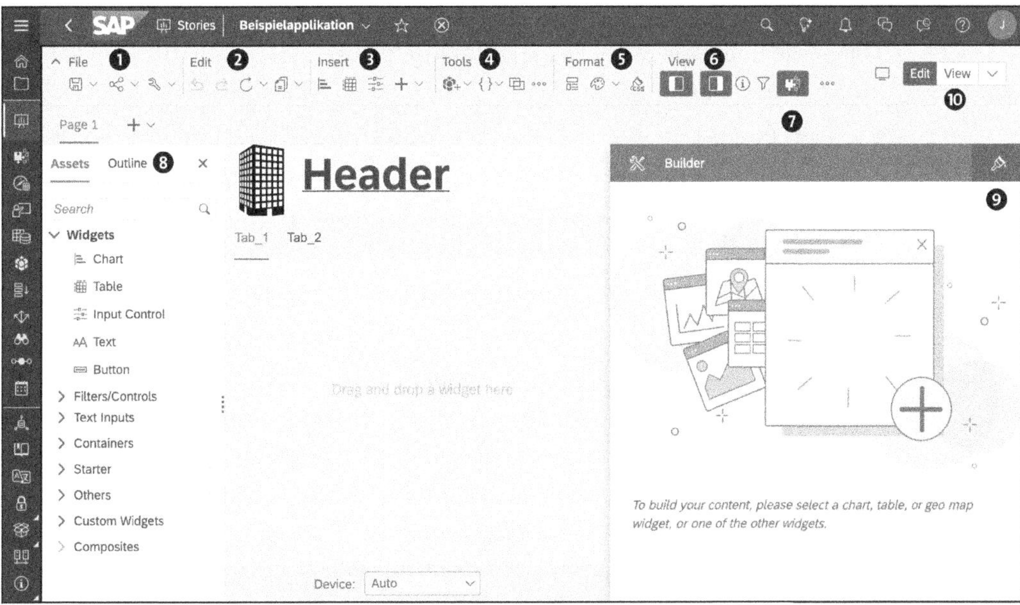

Figure 2.9 Overview of Development Environment

The **Edit** area ❷ offers you the options of resetting or repeating your changes, updating your data, and copying and pasting your objects.

In the **Insert** area ❸, you can insert all available widgets and other elements. A *widget* is an element that you can use to visualize your data, and there are other elements that you can use to do things like structuring.

In the **Tools** area ❹, you can manage the models you use by adding, replacing, or removing models. You can also manage the variables contained here and perform linked analyses, and you have the option of defining refinements to your story, such as by adding scaling and conditional formatting and displaying a diagram of the linked widgets.

In the **Format** area ❺, you can choose from various layouts by selecting an already created theme (Section 2.7.2), or you can create a theme. You can also create a cascading style sheet (CSS) script.

In the **View** area ❻, you can show and hide the left and right side panels as well as the information area, which displays error messages and a reference list. The counter indicates the number of errors, and if you select an error in the info area, you'll be taken to the corresponding faulty script. With the reference list, on the other hand, you can quickly check where an element is used, and this makes it easier for you to understand existing scripts and can help you to check whether you want to delete an element.

In the right side panel ❼, you'll find general functions for your selected element, as well as the visual formatting. This page area is also divided into two sections:

- **Builder**
  This contains general functions, such as the chart type, selected key figures, add-ons, tab settings, and many more.
- **Styling**
  This area lets you define formatting settings such as size, font, color, and alignment.

In the left side panel ❽, under **Assets**, you'll see an overview of the available widgets, which you can drag and drop into the graphics area. In the **Outline** area, you have an overview of the elements used in your story, as well as options for settings that you're familiar with from the **File** area. In the **Pages** subitem, you can see the widgets used and the popups created, and the elements built into another element are displayed indented for better clarity.

You also have the option of renaming individual objects or showing and hiding them. This helps you to maintain a better overview, especially with more complex stories. You can also display references for your object here to help you better assess dependencies. You can use the **fx** button to store code for the specific event of the widget, but you must first select the event for which you want to store code. A new tab then opens in the graphics area in which the code is stored.

In the **Scripting** subitem, you'll find all the script elements that you can create within your story.

You can use the arrow icon to collapse or expand all nodes and the **+** icon to create new popups or script elements.

The graphics area ❾ is your workspace for creating a story. You can customize the graphics area using the format settings in the right side panel, and you can create it dynamically or in a fixed size.

In the upper part of the graphics area, you can use the ⌄ icon to switch between the graphics area and the popups. You can also write the code for your functions there; a tab opens within the graphics area for this purpose. To make writing code easier, you can call up an input help dialog with the key combination [Ctrl]+[Spacebar]. This provides you with a list of suggested APIs from which to choose.

You can switch to view mode by clicking the **View** button ❿, which will open your story in a new browser window. To do this, you must have saved your story at least once; otherwise, the button is grayed out. If you've made changes to your story since the last time you saved it, the last saved status will be executed. If you've made an unsaved change to the story, the name of the story is marked with an asterisk.

[«]

Refactoring

The SAP Analytics Cloud development environment features *automatic refactoring*. This means that when an entity is renamed, all references to this entity (such as scripts, dynamic texts, and calculated key figures) are automatically renamed too. It also means that the application remains intact even after renaming and you don't have to manually adapt all entities.

2.7.2 Design of a Story

In this section, we look at the possible design settings for a story.

Layouts Visual design of an application

For stories with scripting, you can use the existing layout templates that you're already familiar with from simple stories, but when creating the visual design of a story, you have the option of creating your own themes. You can apply these themes to your story or customize them using the corresponding API.
In these themes, you can make overarching settings for various widgets, which makes it easier for you to create a uniform story design.

To create a new theme, go to the editing mode of an existing story or create a new story. In the **Format** area, you can create a new theme using the color palette icon, selecting and editing an existing theme or searching your tenant for an additional theme (as shown in Figure 2.10).

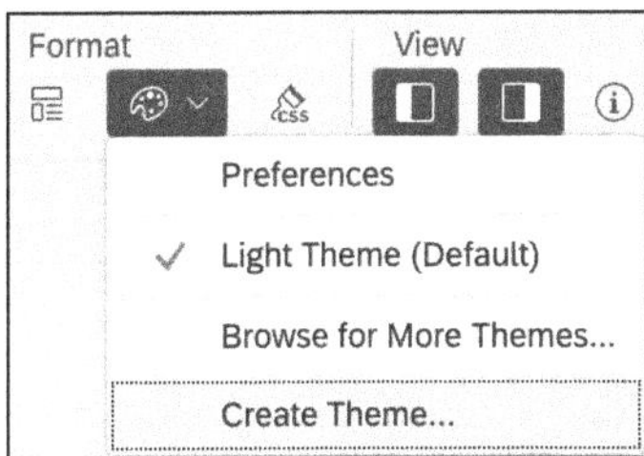

Figure 2.10 Creating New Theme

Create and customize theme

Once you've selected the **Create Theme...** option, a new window will appear, and you can then define your theme preferences for the various widgets in the window (see Figure 2.11). Enter the name of the theme, specify the background color for the graphics area and the popups, and save your design to the desired location in the SAP Analytics Cloud file system by clicking the **Save As** button.

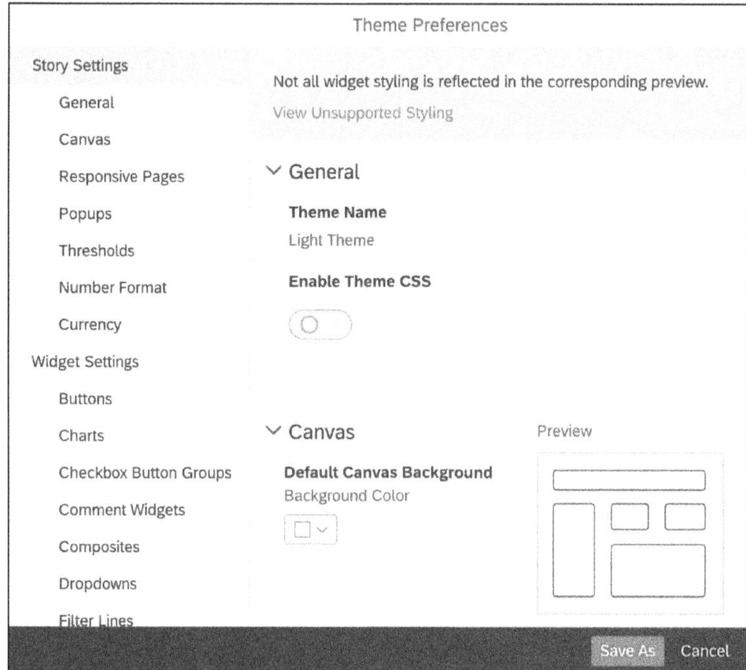

Figure 2.11 Preference Options for Your Theme

You can now use this theme in other stories, and changes to this theme will be transferred to those stories.

Using CSS scripts

In addition to the basic theme preferences, you can store CSS scripts for your widgets. The CSS script options are currently still very limited, and not all CSS properties are available. With a CSS script, you have the option of customizing your widgets.

Page 1 + Story - CSS

Story CSS
Here you can edit CSS for your story.

Button — Note: Select an object from this dropdown list to view its supported class names and properties.

```
/**
* Support Class Names:
*
*    Widget CSS Class:
*        Description: The CSS class you are going to assign to a widget in its styling panel should not be a predefined class.;
*        Properties: background-color, border, border-top, border-right, border-bottom, border-left, opacity;
*        Descendants: N/A;
*        Pseudos: N/A;
*
*    .sap-custom-button-widget:
*        Description: The default class for this kind of widget. It is defined after a widget CSS name but its rules will have higher priority.;
*        Properties: background-color, border, border-top, border-right, border-bottom, border-left, opacity;
*        Descendants: N/A;
```

```
.myButtonBorder .sap-custom-panel-widget {
    border: 3px solid rgba(12,100,140,0.9)
}
```

Figure 2.12 Editing CSS Script

To edit a CSS script to your story, click the brush icon next to the color palette icon that you clicked to create the theme. A tab will open where you can store your CSS script in the desired widgets, and you'll also find information on which properties are available (see Figure 2.12).

You can now store the scripted CSS classes in your widget via the designer (see Figure 2.13).

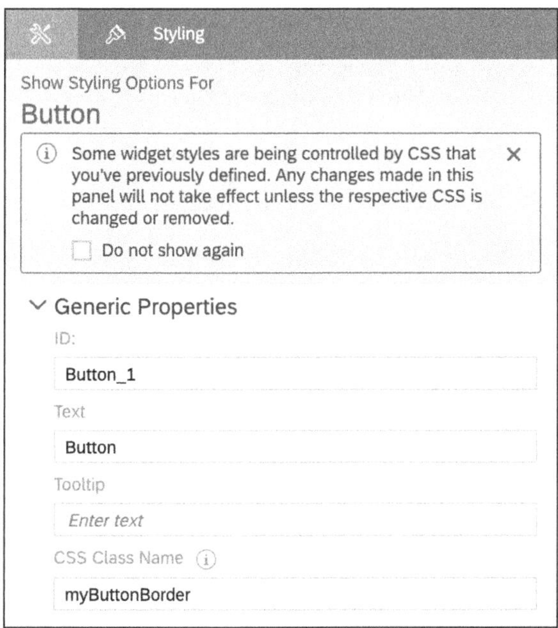

Figure 2.13 Applying CSS Script

2.7.3 Keyboard Shortcuts for Input Help

Important shortcuts

The SAP Analytics Cloud script editor provides you with some helpful keyboard shortcuts that make your daily work easier. These are explained in Table 2.1.

Key or Key Combination	Description
Ctrl + Spacebar	Use this key combination to open the input help in the script editor with suggestions.
Ctrl + A	Use this key combination to select the entire content of the editor.

Table 2.1 Most Important Keyboard Shortcuts for Stories

Key or Key Combination	Description
Ctrl+D	Use this key combination to delete the entire line in which the cursor is located, including the line break character at the end.
Ctrl+S	Use this key combination to save the entire application.
Esc	Use this key to deselect all selections, except the last selection if there's a multiple selection.
Ctrl+Z	Use this key combination to undo the last change.
Ctrl+Y	Use this key combination to restore the last undone change.
Ctrl+U	Use this key combination to undo the last change to the selection (or the last change if the history doesn't contain any changes to the selection).
Alt+U	Use this key combination to re-execute the last undone change to the selection (or the last change to the text if there have been no changes to the selection).
Ctrl+Pos1	Use this key combination to move the cursor to the beginning of the document.
Ctrl+End	Use this key combination to move the cursor to the end of the document.
Alt+←	Use this key combination to move the cursor to the beginning of the line.
Pos1	Use this key to place the cursor at the start of the text in the line, or if it's already there, at the actual start of the line (including spaces).
Alt+→	Use this key combination to move the cursor to the end of the line.
Ctrl+←	Use this key combination to jump one group to the left of the cursor. A *group* is a string of characters, a sequence of punctuation marks, a new line, or a sequence of at least two spaces.
Ctrl+→	Use this key combination to jump one group to the right of the cursor.

Table 2.1 Most Important Keyboard Shortcuts for Stories (Cont.)

Key or Key Combination	Description
Shift+Backspace	Use this key combination to delete the character in front of the cursor.
Delete	Use this key to delete the character after the cursor.
Ctrl+Backspace	Use this key combination to delete the group to the left of the cursor.
Ctrl+Delete	Use this key combination to delete the group to the right of the cursor.
Shift+Tab	Use this key combination to automatically indent the current line or selection.
Ctrl+F	Use this key combination to perform a search.
Ctrl+G	Use this key combination to continue searching.
Shift+Ctrl+G	Use this key combination to search for previous items.
Ctrl+/	Use this key combination to switch between commenting and removing comments from the selected lines.

Table 2.1 Most Important Keyboard Shortcuts for Stories (Cont.)

2.8 Authorization for Scripting with SAP Analytics Cloud

All SAP Analytics Cloud licenses include the creation and use of scripting. For planning, however, you should note that the corresponding planning license must also be active.

Necessary authorizations

To be able to open a story in SAP Analytics Cloud, you need read permission for the story and at least one of the standard BI roles. To use advanced mode, you also need the *application creator* role. If you want to edit a story, you need authorization to edit the corresponding story in addition to the application creator role.

Create your own roles

Of course, you can also integrate these authorizations into your own authorizations. To do this, you create your own role in your SAP Analytics Cloud tenant, and you can define the license type to which your own role should belong. You can see an example of how to create a new role in Figure 2.14.

In the next step, you can use one of the existing roles as a template for the newly created role or start with an empty template and adapt it to your requirements.

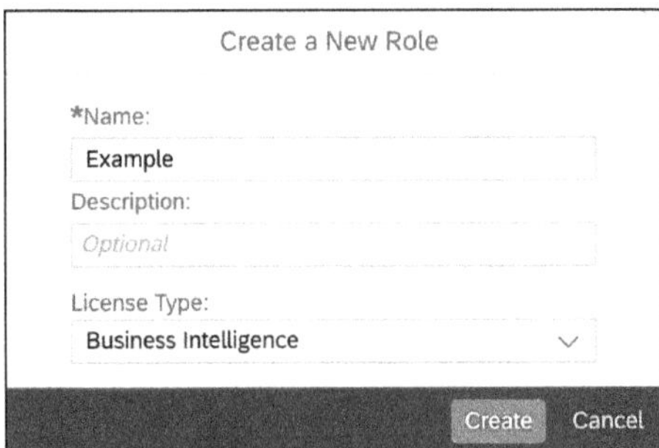

Figure 2.14 Creating Your Own Business Intelligence Role

The standard SAP role is sufficient to execute the examples in this book. However, if you want to work out the role in more detail (e.g., if certain user groups should create scripting but no or only certain widgets), then you must also assign authorizations for stories and user-defined widgets to the role in question. In the case of user-defined widgets, this authorization can take place at the widget level.

Configure role

You can specify whether the corresponding role should be able to create, read, update, delete, or share an object. In Figure 2.15, we have highlighted the lines for stories and user-defined widgets in color. Select the appropriate settings for your role here.

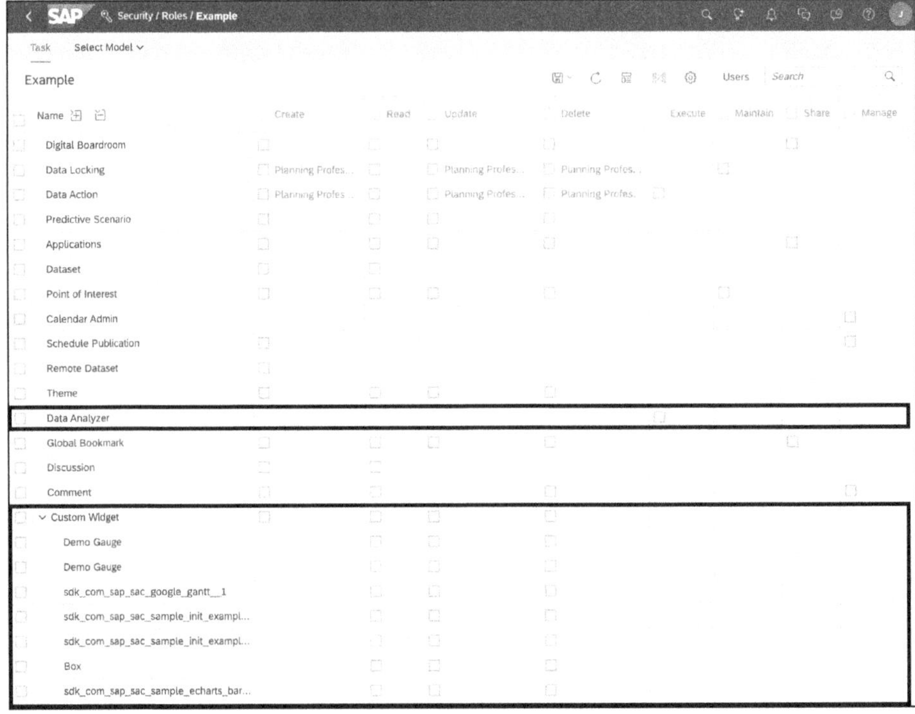

Figure 2.15 Setting Options for Role

Import of demo applications

To start you off in advanced mode, SAP provides you with various demo stories. These are not available in your SAP Analytics Cloud tenant right from the start but must be imported.

Please note that you'll often still see stories in Analytics Designer, which is the predecessor of advanced mode. However, this makes no difference to you in terms of functionality, and we'll show you how to convert the old elements of Analytics Designer into a story in advanced mode later in this chapter.

Here are the steps you need to take to import the demos:

1. **Open the content network**
   Open your SAP Analytics Cloud tenant and switch to the Content **Network** area (see Figure 2.16).

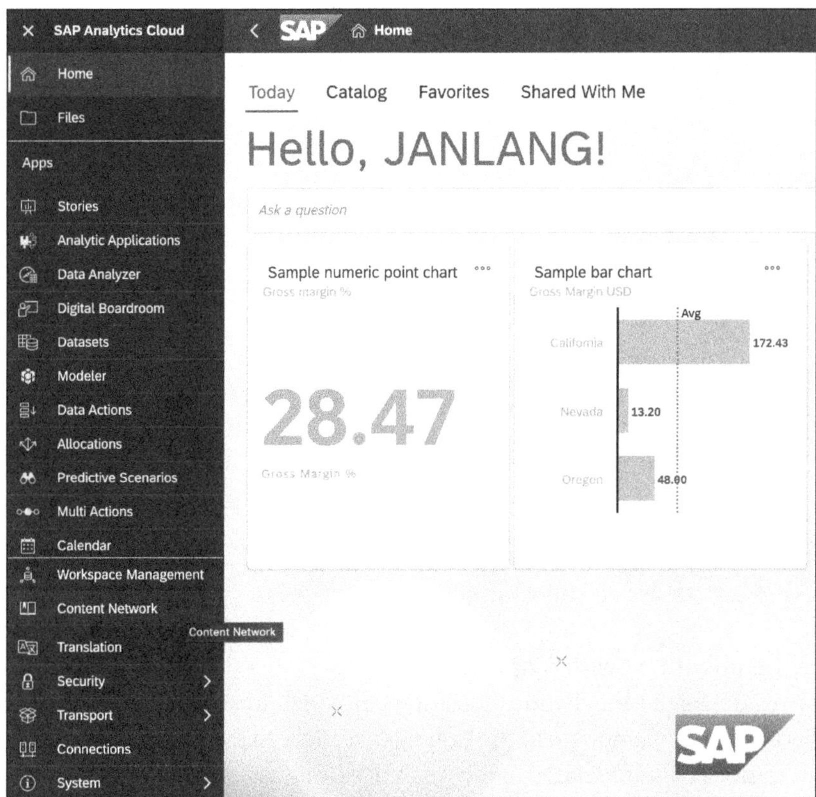

Figure 2.16 Opening Content Network

Differences in demo content

2. **Review your options in the content network**
   In the content network, you'll find various contents that you can import into your SAP Analytics Cloud tenant. These are divided into three areas (see Figure 2.17):

- **Samples**
  In the **Samples** section, you'll find sample content for the various components of SAP Analytics Cloud. Among other things, you'll find templates and the well-known best-run examples. You can also find examples of the Analytics Designer here.
- **Business Content**
  SAP provides **Business Content** and offers ready-made dashboards for different industries and sectors on which you can base your dashboards.
- **3rd Party Business Content**
  3rd Party Business Content is similar to **Business Content**, except that it is not created by SAP. Here, SAP partners have the opportunity to make their developed content available.

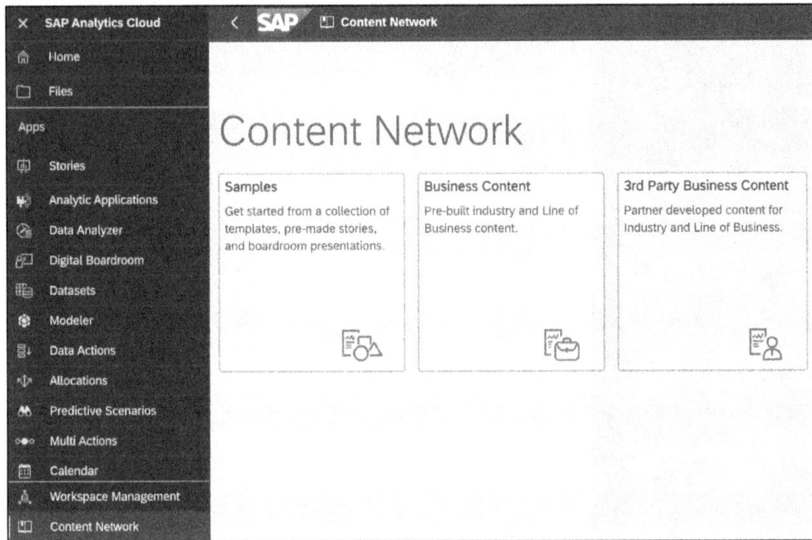

Figure 2.17 Areas of Content Network

3. **Demo content with scripting**
   To download the demo applications provided by SAP for the Analytics Designer, you now go to the **Examples** section. In this section, you'll find various content related to the Analytics Designer:
   - **Sample Analytics Application for Mobile**
     This package shows you how to design applications for mobile devices.
   - **Analytics Designer**
     This package provides you with a demo application in which you can use the layout API. Specific examples of various API functions are also available here.

- **Analytics Designer - Ready-to-Run Applications**
  This package provides you with complete applications.

In Figure 2.18, you can see an overview of the available content.

SAP Content Network / **Samples**

Name	Description	Created By
SAP Analytics Cloud Accessibility Tem...	Content SAP Analytics Cloud Accessibility...	SAP
SAP Sample Content for FI, HR and SD	SAP Sample Content for Finance, Human ...	SAP
SAP Analytics Cloud Usage Tracking C...	Please use the new System Overview!	SAP
Sample Content for Planning	This package covers SAP Analytics Cloud ...	SAP
Sample Analytics Application for Mobile	Content Sample Analytics Application for ...	SAP
Sample Content (BestRun Juice Com...	Sample Content (BestRun Juice Company)	SAP
SAP Analytics Cloud Guidelines Templ...	Content for SAP Analytics Cloud Template...	SAP
Analytics Designer - Ready-to-Run Ap...	SAP Analytics Cloud, Ready-to-Run Analy...	SAP
Analytics Designer	SAP Analytics Cloud, Analytic Designer Sa...	SAP
SAC BestRun Bikes Mobile demo	SAC BestRun Bikes Mobile demo	SAP
SAP Samples What-If Application	Content for SAP Samples What-If Applicat...	SAP
My Team Dashboard	My Team Dashbord Sample	
Management Dashboard	JF Technologies	SAP

Figure 2.18 Overview of Examples of Content Network

4. **Prepare to import a package from the content network**
   To download a package from the content network, first select the desired package. In our example, we are importing the **Analytics Designer** package.

   After you select the package, a popup opens containing the two tabs: **Overview** and **Import Overview**. The **Overview** tab is in turn divided into the following areas (see Figure 2.19):

   - **Import Overview**
     Here, you can see which settings you've made in the import options for your import.
   - **Description**
     This section gives you an overview of what this package is about.
   - **Details**
     This section lists in detail what content is included in this package; in our example, it lists the individual features.
   - **Created by**
     In this column, you can see who created this package. This is particularly relevant in the area of third-party content.

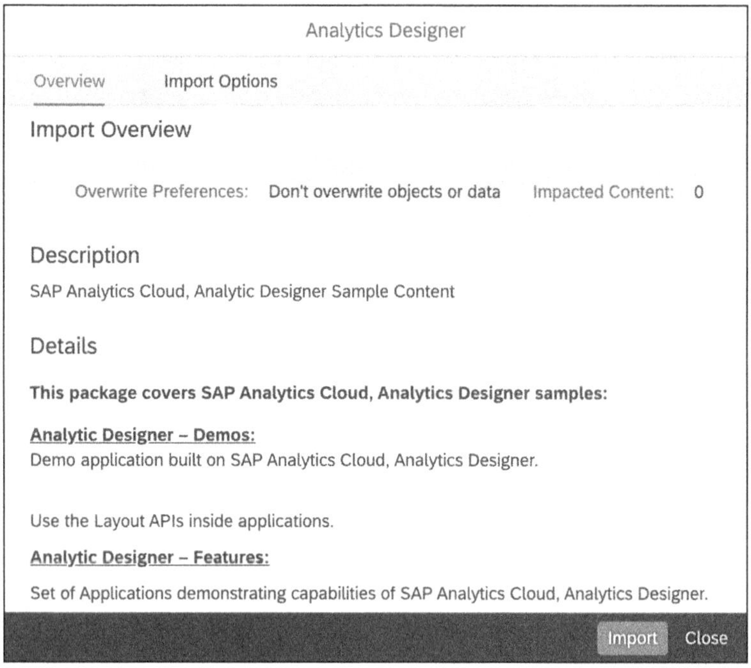

Analytics Designer
Overview Import Options
Import Overview
Overwrite Preferences: Don't overwrite objects or data Impacted Content: 0
Description
SAP Analytics Cloud, Analytic Designer Sample Content
Details
This package covers SAP Analytics Cloud, Analytics Designer samples:
Analytic Designer – Demos:
Demo application built on SAP Analytics Cloud, Analytics Designer.
Use the Layout APIs inside applications.
Analytic Designer – Features:
Set of Applications demonstrating capabilities of SAP Analytics Cloud, Analytics Designer.
Import Close

Figure 2.19 Overview of Package

The **Import Options** tab is divided into three areas (see Figure 2.20):

- **Import Schedule**
  In this area, you can set your own schedule. However, this option must be offered by the content creator.
- **Overwrite Preferences**
  In this area, you can specify whether the selected package may overwrite objects and data or whether existing content should not be overwritten.
- **Content**
  This area provides you with very detailed information about which objects are imported into your SAP Analytics Cloud tenant during an import.

5. **Import the package**
   You can now start the import by clicking the **Import** button, which causes the import to be carried out in the background. As soon as the import has been successfully completed, you'll receive a notification in your notification area, where you can also track the current status of the import at any time.

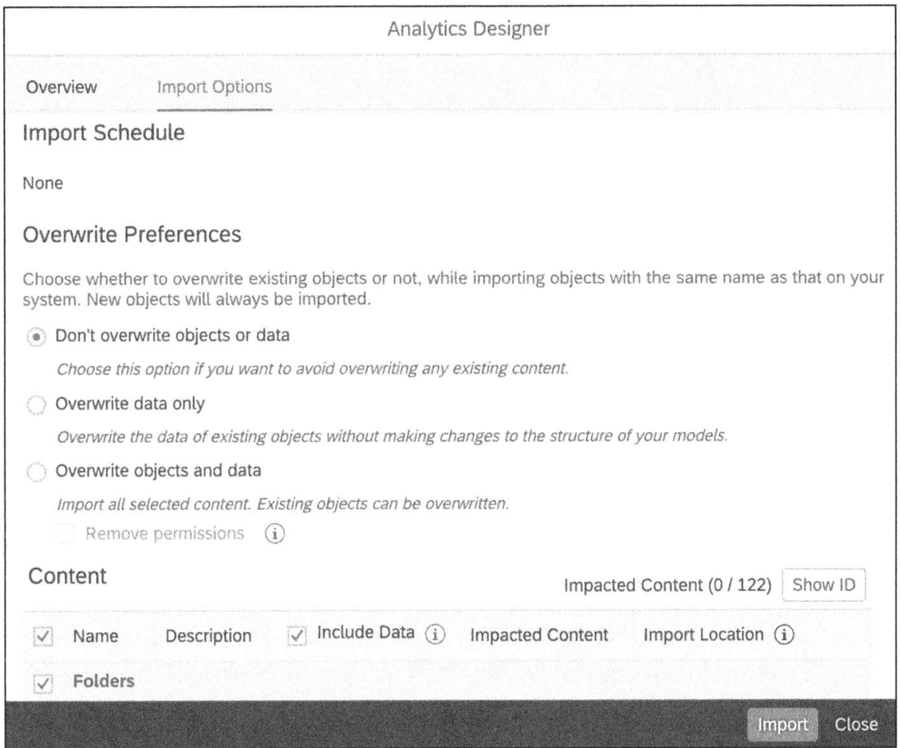

Figure 2.20 Importing Overview of Package

2.9 Conversion of a Classic Story or Analytic Application

Up to this point, you've learned how to create a new story with advanced mode, but what should you do if you've already created classic stories or analytic applications? How do you get them into the new optimized design experience?

Don't worry, SAP has thought of this case and developed migration options. We present them to you here, as you can see in Figure 2.21.

How you can transfer your existing classic stories and analytic applications depends on their current status. For this purpose, the *optimized view mode* was introduced as an intermediate step. This is a basic prerequisite for transferring your classic objects to the *optimized design experience*.

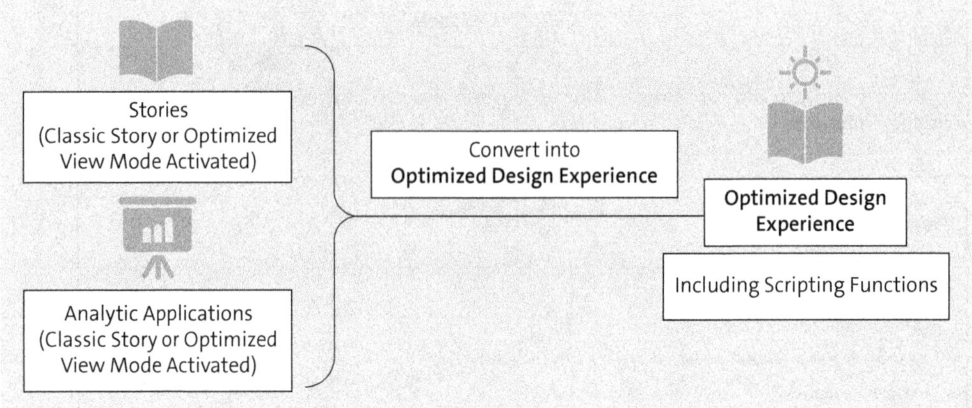

Figure 2.21 Converting into Optimized Design Experience

Stories that were built in or converted to the optimized design experience will be automatically converted to the optimized story experience in the future. The same applies to stories and analytical applications for which the optimized view mode is activated. In the future, these will also be opened automatically in the runtime environment of the optimized story experience.

The optimized mode can be set in the details of the story or analytic application (see the **Enable Optimized Mode** switch in Figure 2.22).

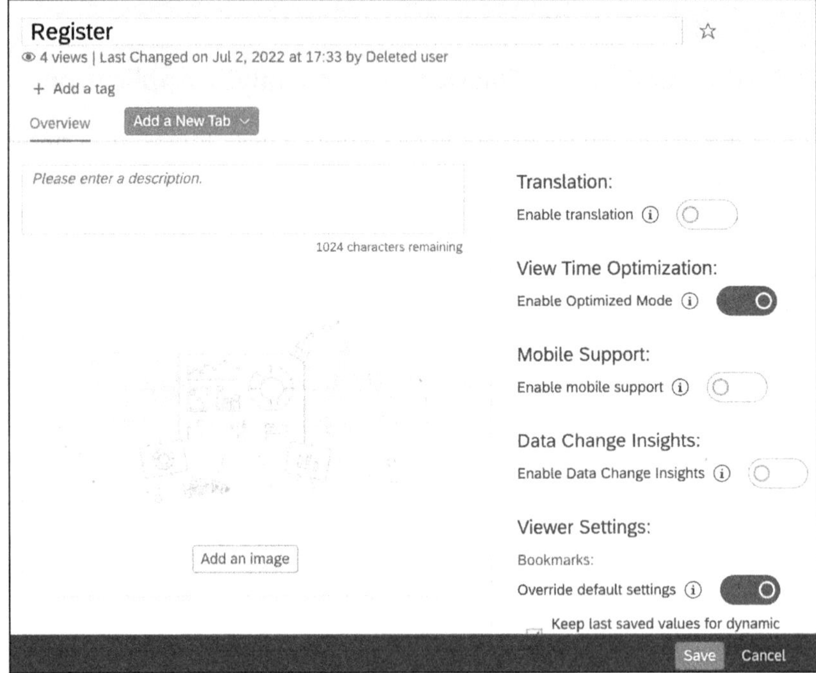

Figure 2.22 Activating Optimized View Mode

If your story or application isn't automatically converted to the optimized design experience, you can carry out these steps yourself, provided that all requirements are met.

First, open your existing classic story or application. In the **Save** menu, you'll see the additional subitem **Convert to Optimized Design Experience**. This is a one-way conversion and can't be undone (see Figure 2.23).

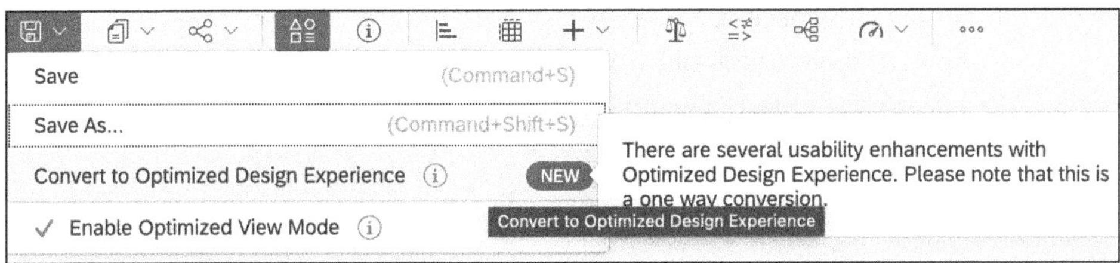

Figure 2.23 Converting to Optimized Design Experience

You can now convert your object. A warning will first appear here, describing all the resulting changes and effects. You can confirm the conversion by clicking the **Convert button** (see Figure 2.24).

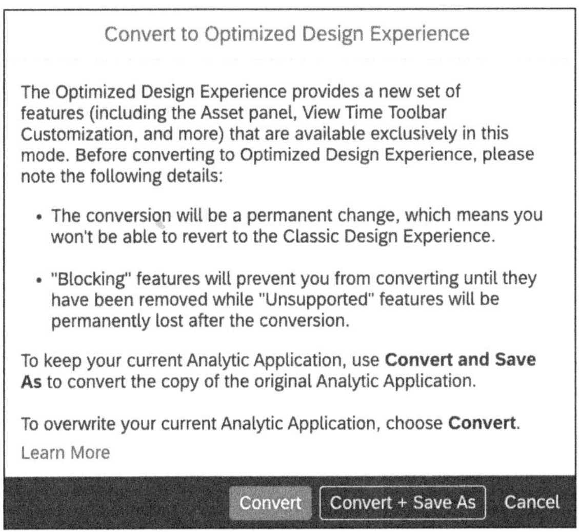

Figure 2.24 Saving Optimized Design Experience

You complete the conversion by selecting the storage location and clicking **Save**. At that point, you can benefit from all the innovations of the new design experience in your old system.

2.10 Summary

In this chapter, you've learned about the main aspects of a potential system landscape and the basics of data connections and modeling. Furthermore, you've learned the difference between a classic story and a story with scripting from different points of view. You've gotten to know the development environment and learned how to get started with the business content and the necessary system rights. At the end of the chapter, you learned how to convert your old stories into the new optimized design experience.

In the next chapter, we get started with scripting!

3

Chapter 3
Development of Business Intelligence Applications

In this chapter, you'll learn about the various widgets and APIs that you can use in stories and your scripts in SAP Analytics Cloud. We'll also look at the typical elements of a BI story.

What you can expect

So far, you've learned the basics of SAP Analytics Cloud, and we've dealt with the integrated development environment and the extended mode. Now, we'll dig deeper into widgets and APIs. In Section 3.1, "Widgets," we'll explore existing widgets and other widgets that are particularly suitable for extensions with scripting. In Section 3.2, "Application Programming Interfaces," we'll explain the most important APIs. In Section 3.3, "Developing a Story with Scripting for Reporting," we'll show you sample scripting scenarios so that you can understand the lifecycle of a story extended by scripting. In Section 3.4, "Integration of Predictive Analytics Functions," we'll look at the predictive functionalities that you can use in your story. Finally, in Section 3.5, "Best Practices," we'll present some best practices for scripting.

3.1 Widgets

Widgets in an application

There are *widgets* that are particularly suitable for scripting because they give a story structure and flexibility, which allows you to make your stories more interactive. You will see that there are several different widgets available with which you can map similar use cases. Here, which widget you use is a question of design.

Selecting Widgets

Which widget is most suitable for your needs often depends on your corporate design and what other software solutions your company uses. For example, you should not exclusively use checkboxes in your BI stories if dropdown lists are used in all your other software solutions. Try to put yourself in the end users' shoes to provide them with the best possible user experience. This will increase user acceptance of your BI stories.

To insert a widget into your story, create a new story or open an existing one. In the development environment, you'll find the add [+ ⌄] icon, which you can use to add different widgets to the graphics area (see Figure 3.1). We'll take a look at these widgets in the following sections.

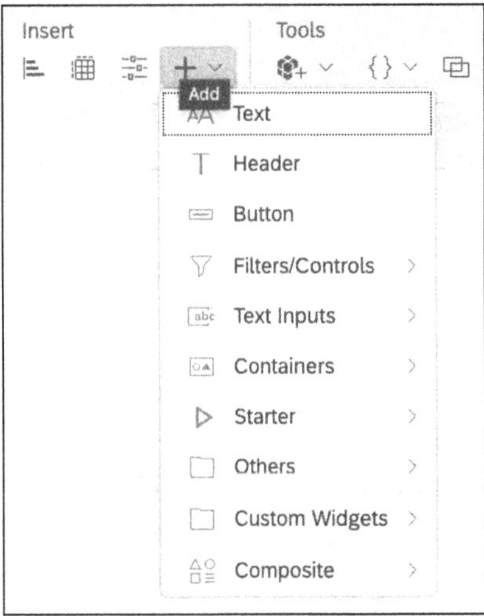

Figure 3.1 Inserting Widget

3.1.1 Panel Widgets

Divide into areas

Panel widgets are particularly useful for organizing and structuring your story. With a panel widget, you can group different elements together and nest several elements inside each other. This allows you to divide your story into different areas and allocate a certain amount of screen space to them via the right-hand side area. The elements within the area are also assigned a defined size, and that allows you to quickly and easily move areas, show and hide them, and quickly adapt your story to new screen sizes.

For example, you can create a panel in your story for a header ❶ that always takes up 10% of the screen space (see Figure 3.2). There's also a main panel ❷ that takes up the remaining 90% of the screen, and you can divide this main panel into a filter area ❸ and a display panel ❹ with further panel elements.

[»]

Preliminary Considerations

As early as possible, start thinking about how you want your story to look and which areas you need to include to make it look that way. You can move widgets from one area to another at any time, but the size and position settings are often lost in the process, so it's best to determine the layout before creating your story.

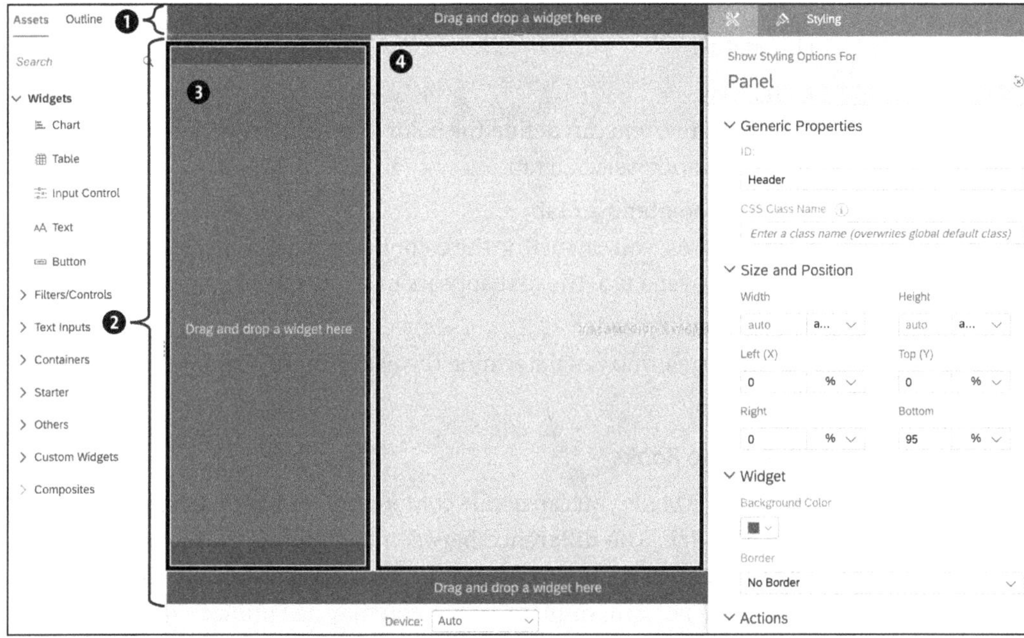

Figure 3.2 Nested Areas

3.1.2 Tab Strips

Subdivision with tabs

With the help of a *tab strip*, you can divide your story into different tabs. This helps you separate the content of the story thematically. By default, the initial tab contains two cards, and you can use the builder on the right-hand side of the page to add further tabs and select a standard tab to be displayed first. You can also customize the names of the tabs here and assign a technical name with additional display text that is displayed in the individual tabs. You can see what this looks like in SAP Analytics Cloud in Figure 3.3.

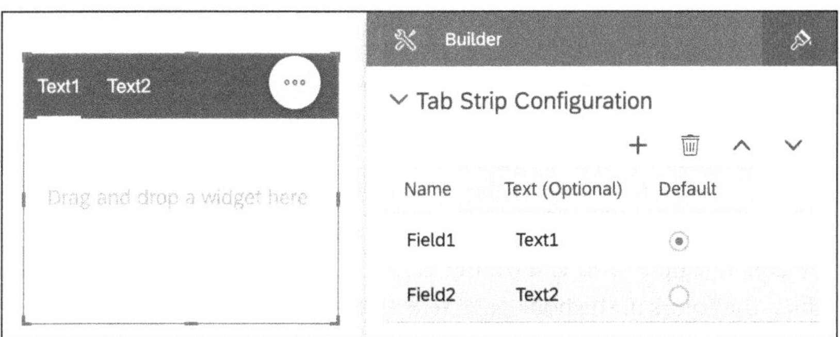

Figure 3.3 Creating Settings for Register

In the **Format** area, you can adjust the style of the tab depending on the specific status:

- **Selected**
  In this area, you can define the color of the selection bar and the font of the currently selected tab.
- **Mouse pointer over tab**
  In this area, you can define the color that appears when a user moves the mouse over a tab. The text appears in the specified color.
- **Background header**
  In this area, you can determine the background color of the entire tab.

3.1.3 Page Books

Book pages in a story

With a *page book*, you can divide content into different areas, as you would do with a tab. The difference between page books and tabs lies in their appearance. There are no tabs at the top of a page book, but there are small dots at the bottom, similar to what you may be familiar with from your e-book reader (see Figure 3.4). You can use these dots to switch among different book pages.

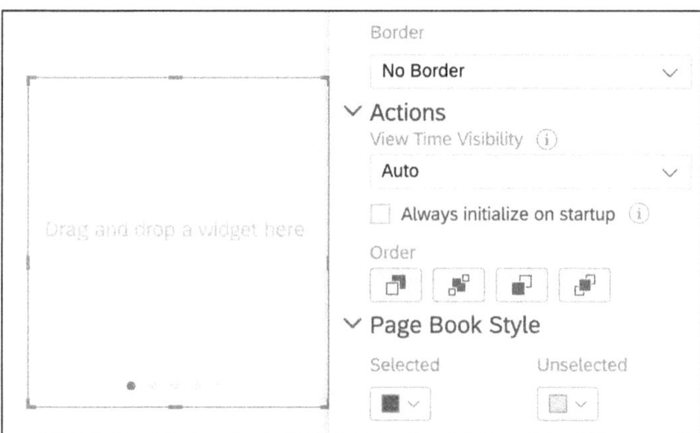

Figure 3.4 Setting Colors of Book Page

By default, a page book consists of two book pages, but you can use the builder to add pages and specify a default page. As no name is displayed in the element here, you can't enter any text. In the **Format** area, you can adjust the colors of the book page selection area, for example.

3.1.4 Flow Layout Panel

Flexible stories with the flow panel

The *flow layout panel* widget ensures that all included elements are displayed in their optimal position, regardless of screen size or device (see Figure 3.5). For example, if the screen is too small to display two contained elements next to each other, the widget moves the second element to the next line.

You can also use the builder to create rules in the form of break points. This allows you to determine how the story should react when screen sizes fall below certain limits. In this way, you can reduce the size of included widgets or hide them.

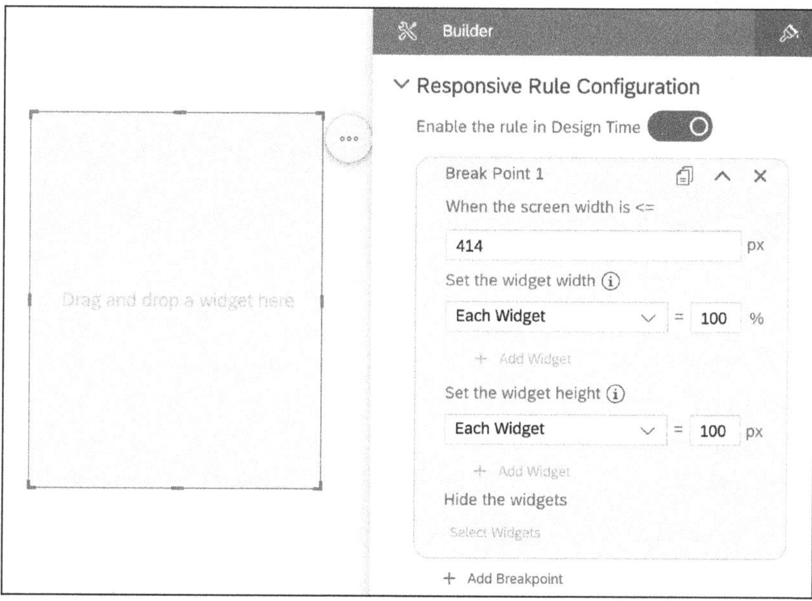

Figure 3.5 Configuring Flexible Rules in Flow Layout Area

[+]

Create the Right Look and Feel

By combining the flow layout area and a simple area, you can very quickly and easily imitate the look and feel of the SAP Fiori launchpad. You can find an example of this in the demo content, and we'll go into this in Section 3.3.2 in more detail.

3.1.5 Input Field and Text Area

Make text input possible

Since the *input field* and *text area* widgets are very similar in their function, we present them together here.

With both elements, you give your end user the option of entering their own values (e.g., an order number). You can then process these values further in your story. In the builder, you can also define default values that are displayed the first time they are called up. There are five different options here:

- **Manual input**
  With manual input, you can enter a value yourself. Here, you can send a request to the user, such as "Please enter an order number."
- **Script variables**
  With a script variable, you can use the value of a defined variable that you've created in your story.
- **Model variables**
  With a model variable, you can select a variable of a model you use.
- **Tile filters and variables**
  With tile filters and variables, you can select a set filter for a tile that you use in your story.
- **Properties of the story**
  You can choose between different properties of the story:
  - Current user or users
  - Current time
  - Current date
  - Date of last change
  - Date and time of last change
  - Last modified by
  - Creator or creators

You can also specify whether the value entered by the user should be written back to a variable in the story.

Difference between an Input Field and a Text Area

The difference between an input field and a text area is the layout of the inserted characters. In an input field, the text is always displayed on one line, which means that if an entered text is too long to be displayed, it will be cut off. Therefore, for longer texts, you should use a text area, which will automatically insert line breaks to keep the text complete and legible. You can also insert line breaks manually during input, as we show in Figure 3.6.

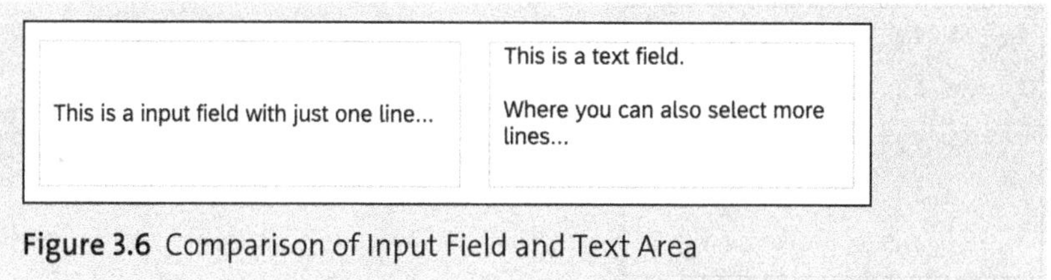

Figure 3.6 Comparison of Input Field and Text Area

3.1.6 Dropdown Lists

Selection with dropdowns

With *dropdown lists*, you can make your story more dynamic by giving users the option of selecting values themselves (see Figure 3.7). They can, for example, display a key figure or search for a specific year using a filter.

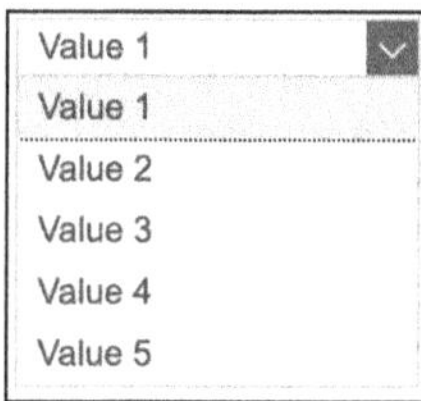

Figure 3.7 Selection in Dropdown List

You can manually define the values that are displayed in the dropdown list or read them in via various variables. You already know these variables from the input field widget, and only the properties of the story can't be selected here. You can also adjust the format and specify different colors for the dropdown list. You can define the frame and background of the dropdown list here, and as you know from the tab, you can select different colors for the selected field, a mouseover, or holding down a mouse button.

3.1.7 Checkbox Groups and Radio Button Groups

Various selection options

Both *checkbox groups* and *radio button groups* are very well suited to giving end users the option of selecting one or more values, but there is a big difference between these two widgets. With a checkbox group, the end user has the choice of selecting several values from the defined set, while with a radio button, the end user can select only one value. These widgets are therefore also similar to the dropdown list widget described above, and you can see the different displays of the two widgets in Figure 3.8.

Figure 3.8 Difference between Checkbox Group and Radio Button Group

3.1.8 Buttons

Triggering actions with the buttons

A *button* is a very variable element that you can use for many different purposes. Basically, clicking a button can trigger an action in a story, and you can enter this action as text in the button so that it's clear to users what the click will do (see Figure 3.9). In addition to the text, you can upload an icon and have it displayed on the button. You can also integrate mouseover text here.

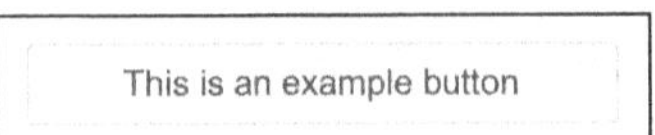

Figure 3.9 View of Button

Possible applications for a button include the following:

- Calling up popups
- Switching between different widgets
- Exporting the contents of certain fields
- Sending emails

You can adjust the color design of your button depending on the status of the button, and you can change the frame and background depending on whether the user presses the button or hovers a mouse cursor over it. You're also provided with five predefined design types (e.g., a positive button, which is intended as an acceptance button).

3.1.9 Simple Sliders and Range Sliders

Flexibility with controllers

You can use *simple sliders* and *range sliders* to flexibly enter values (e.g., months, amounts of money) into a story.

A simple slider can be used to select a fixed value, such as a 10% discount. The end user can then use this slider to simulate the effects of a discount on a potential purchase.

A range slider, on the other hand, is used to select a range of values. For example, a user can set a range slider to only display products with a discount of 30% to 70 % (see Figure 3.10).

Figure 3.10 Difference between Simple Slider and Range Slider

As with the dropdown list in the right-side panel, you can read in the values of the two sliders manually or via different variables. You can also specify whether the identifiers should be displayed and how the values should be changed (via drag and drop and/or direct inputs). You can also optionally specify a step size and adjust the colors of the progress bar and background bar.

3.1.10 Filter Line

Filter by important data

The *filter line* in a story is very similar to the familiar filters from the story, except that you can use it to choose which widgets are affected.

The source of the filter line can be defined in two different ways:

- **Group filter**
  In this mode, you can select a data source that you use in your story, and you can define a dimension selection and predefined filters from it. You can then apply these to the entire data source or only to certain widgets.
- **Individual widget filter**
  In this mode, you select a widget that serves as a source, and you can then define a dimension selection from this source widget.

You can also customize the color scheme of the filter line in the designer in the corresponding company design. In Figure 3.11, you can see how a filter bar is displayed.

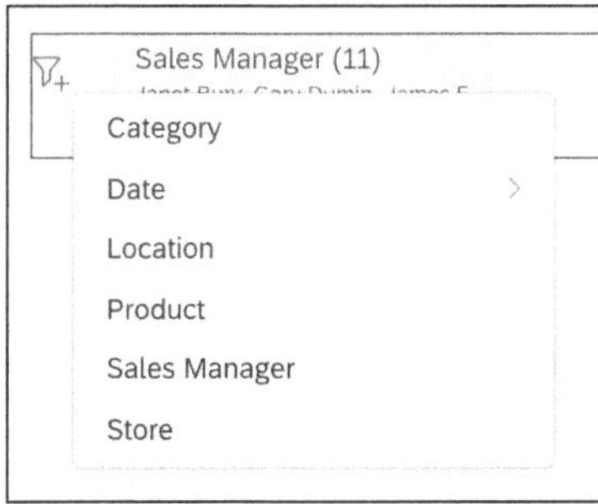

Figure 3.11 Insert Filter Line in Group Filter Mode

3.1.11 List Boxes

Lists in a story

A *list box* is similar to a dropdown list, a checkbox group, or a radio button group and is used in similar application scenarios. You can either store manual values in list fields or read them in via a variable. You can assign a technical ID and the name to be displayed, and you can also specify how a text overflow should be handled. Furthermore, you can specify whether you want to enable multiple selection, activate HTML tags, or write the selected values back to a variable at runtime.

You have the same selection options for the color design of a list box as you have for a dropdown list. You can see the implementation of an example list field in Figure 3.12.

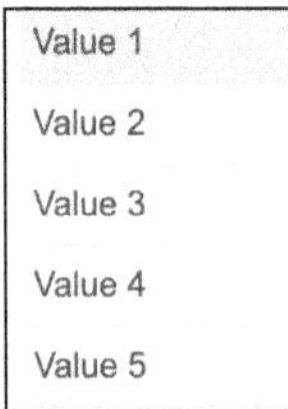

Figure 3.12 Display of List

3.1.12 Switch

Change status with a switch

The *switch* widget is comparable to a light switch because you can turn it on or off (see Figure 3.13). You define which action is to be executed in the story when the switch is on and when it is off. For example, the switch can control the units in which the story displays time, so when the switch is on, the story can display time in seconds, and while when the switch is off, the story can display time in milliseconds. You can transfer and extend this example to many different application scenarios, and you can set several actions to be carried out during the switchover.

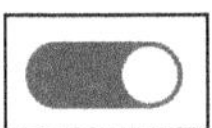

Figure 3.13 Display of Switch

3.1.13 Popups

Create additional space with popups

A popup is not a widget in itself, as it's created in the layout area in your story. From our point of view, however, it fits very well into this section of the book, as it's also an element for displaying data.

A *popup* is an element that you can call up within your story. The application displays the popup when you call it up and hides the graphics area. To create a new popup, go to the **Layout** area in the structure of your application, where you'll find the **Popups** area. All elements of the popup type in your story are listed in this area (see Figure 3.14).

Pages +
Page_1 *Page_1*
Lane_1
Popup_1
Popup_2
Popup_3
Popup_4
Popup_5

Figure 3.14 Organization of Several Popups

Use the **+** icon to create a new popup. The development environment automatically opens the popup you've just created, and you can then insert whatever further elements in this popup you're familiar with from the graphics area. As soon as you've created at least one popup, you can switch between the graphics area and the various popups. In the **Outline** area, you can display this element in the work area by clicking on the desired popup or graphic area.

If you've added a popup to your story, you can make various settings for it using the designer. In the builder area, you have the option of creating a header and footer for your popup. If you don't activate this option, no further setting options will be displayed, but if you activate this option via the switch, the development environment automatically creates a header and footer in your selected popup and the setting options in the builder are extended (see Figure 3.15).

Figure 3.15 Popup Settings of Builder

In the **Title** field, enter a name for your popup that will be displayed in the header. In the **Buttons** area, edit the buttons (which in our example are **OK** and **Cancel)** that are located in the footer. You can also create new buttons here, remove existing buttons, or change the order of the buttons. Furthermore, you can assign an ID and a text for each button and decide whether it should appear highlighted.

You've now created a popup and defined which elements it should contain, and you can make further settings in the **Format** area (see Figure 3.16). In addition to the classic settings that you're familiar with from other widgets, such as the name and the background color, you can define the size of your popup here. Various suggestions are available, or you can define the width and height of the popup yourself.

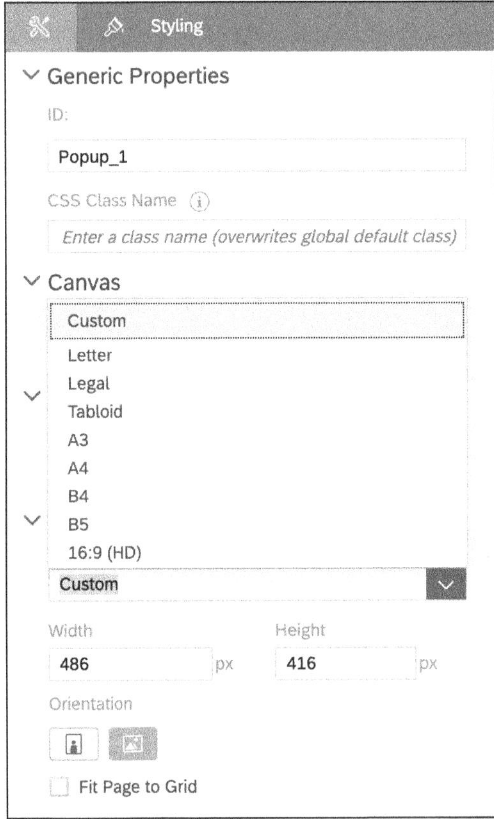

Figure 3.16 Format Settings of Popup

Size of a Popup

The size of a popup is not dynamic, so when creating a popup, please note on which screens your story is to be displayed later.

3.2 Application Programming Interfaces

Develop logic with APIs

In Section 3.1, you learned about the various widgets that are available in a story. However, you've not yet implemented any logic using widgets alone. To set up a story with scripting, you still need *application programming interfaces* (APIs), which are functions that are executed when defined events occur. APIs in SAP Analytics Cloud are not APIs in the classic sense (interfaces with other applications), but rather, they are functions or classes that contain various methods.

We'll now present the SAP Analytics Cloud APIs to you. For a better overview, the individual APIs are divided into different groupings, called *libraries*. Some functions are inherited from one API by another API and can also be used there. We go into these functions in detail in the descriptions of the APIs, and we also show you specific examples of some APIs.

To ensure that this chapter is not too extensive, we explain the most important APIs (from our point of view) in more detail but explain others in less detail. You can find all further information on the remaining APIs and their functions in the "Analytics Designer API Reference Guide" (*http://s-prs.co/v594402*).

[+]

Range of Functions

Depending on the data source you use, some functions may not be available to you or may only be available to a limited extent.

An API can contain the following three different element types, although not every element type has to be present:

- **Events**
  These are actions that are executed
  within your story. An *event* can be anything from a mouse click on an element to the initialization of your story. You can use one or more methods within an event, and you can select events in the **Layout** subitem in the **Outline** area (which you know from Chapter 2, Section 2.6.1).
- **Methods**
  You can execute a method within an event. A *method* contains the specific instructions that you want to carry out within the event, such as hiding a chart or adding a key figure.
- **Properties**
  These are the properties that a method can assume, like the message type of an API if you want to issue a warning.

3.2.1 Standard

Basic APIs

This section covers the APIs of the **Standard** type library, which lists the basic APIs that you need in almost every application.

Application

In the `application` API, you'll find events and methods that affect your story in general. The following events are available in this API:

- **onInitialization**
  The `onInitialization` event is called each time your story is called up when the story loading process has been completed.
- **onOrientationChange**
  The `onOrientationChange` event is called when the orientation of the mobile device is changed.
- **onPostMessageReceived**
  The `onPostMessageReceived` event is called when your story receives a message from the hosting page or an embedded page. However, you should always check the origin of this message, as otherwise, your data sent via the post message could be intercepted.
- **onResize**
  The `onResize` event is called when the window size is adjusted while the story is being used. This event is called at intervals of 500 milliseconds while the window size is being adjusted.
- **onShake**
  The `onShake` event is called when the end user shakes the mobile device. It's only called once within two seconds during the shaking.

Various methods are also available in the API (see Table 3.1).

Method	Description
`getCssClass/setCssClass`	Use this method to read the CSS class from the graphics area or set it in your story.
`getGlobalCssClass/ setGlobalCssClass`	Use this method to read the global CSS class or set it in your story.
`getInfo`	This method sends back information of the `ApplicationInfo` type about your story.
`getInnerHeight/ getInnerWidth`	This method returns the height or width of the graphic area.

Table 3.1 Methods of API Application

Method	Description
getMode	This method returns the mode of the ApplicationMode type of your story.
getTheme/setTheme	Use this method to get back the ID of the current topic or set a topic.
getUserInfo	This method returns information of the UserInfo type about the current SAP Analytics Cloud user.
hideBusyIndicator/ showBusyIndicator	Use this method to show or hide the cross-story busy indicator.
isCommentModeEnabled/ setCommentModeEnabled	This method returns a Boolean value that either indicates whether the comment mode is active or sets it accordingly.
isMobile	This method reports back to you whether your story is in mobile mode.
moveWidget	Use this method to move a widget in your story.
openShareApplicationDialog	Use this method to open the dialog for sharing a story for the currently open story.
postMessage	Use this method to post a message.
refreshData	Use this method to update the data of a data source.
sendNotification	Use this method to send messages. You can find details in the NotificationOptions API.
setAutomaticBusyIndicatorEnabled	Use this method to activate or deactivate the automatic busy indicator.
setMessageTypesToShow	This method specifies which messages are shown, depending on their message type.
setRefreshPaused	Use this method to activate or deactivate the updating of a data source.
showMessage	This method displays a message of the ApplicationMessageType message type in the story.

Table 3.1 Methods of API Application (Cont.)

Listing 3.1 shows you some examples of how you can use the methods of this API.

```
// Displays a success message in your story
Application.showMessage(ApplicationMessageType.Success, "This is a
success message");
// Reads out the user information in your story
Application.getUserInfo();
// Displays the text "This is
// an example"
in the busy indicator of your story
Application.showBusyIndicator("This is an example");
// Refreshes the data of your story
Application.refreshData();
```

Listing 3.1 Examples of Application API Methods

ApplicationInfo

The `ApplicationInfo` API returns the description, ID, or name of your current story. You can see an example of the `ApplicationInfo` API here:

```
// Outputs the description of the application info
 Application.getInfo().description;
```

Console

Check in the console

With the `Console` API, you can display your output as a message in the console of your web browser. This is very helpful for checking the current value of variables during processing. Here, you can see the code for the output of application information and an example variable:

```
// Message output of application information and a
// variable in the console log
console.log(Application.getInfo().name);
console.log(Example variable);
```

Figure 3.17 Output Console Log

You can view this log in the console of your browser (see Figure 3.17). The first line of the log shows you the information read from the name, and the second line shows you the current value of the example variable.

Layout

With the Layout API, you can either read or set the distance between your widget and its parent container. Here, you can access the same values that you know from the designer. There's a get and a set method available for the height, width, and left and right margins. For example, you can use the following code to read out the height of Chart 1:

```
//Reads out the height of Chart 1
Chart_1.getLayout().getHeight();
```

LayoutValue

With the LayoutValue API, you define the unit of your layout, as you know it from the designer. Here, you have a choice of auto, percent, or pixel.

NavigationUtils

The NavigationUtils API is very valuable if you want to call up other elements from your story in SAP Analytics Cloud or if you want to link to external pages.

The following methods are available to you:

- **createStoryURL**
  Use this method to create a URL for a story and have the option of specifying URL parameters.
- **openStory**
  Use this method to open a story with the desired URL parameters and to decide whether it should be opened in a new tab.
- **openDataAnalyzer**
  Use this method to access the data analyzer within your story, specify your desired connection and the URL parameters, and choose whether to open the data analyzer in the same tab or in a new tab.
- **openUrl**
  Use this method to open a URL in the same tab or in a new tab.

The following code shows an example of how to use one of these methods:

```
// Opens the application with the selected ID
NavigationUtils.openStory("5E4B6A0F0E3BFA470DD2BA56044670E9");
```

NotificationOptions

Create notifications

With the `NotificationOptions` API, you can define the content of a notification from SAP Analytics Cloud. You can then send this via the `sendNotification` method of the `Application` API.

You can define the following properties in this API:

- `content`
  Use this property to define the content of your message.
- `isSendEmail`
  This property returns whether the notification is sent by email and/or mobile notification.
- `mode`
  This property specifies the mode in which the story is displayed when it's opened via the notification.
- `parameters`
  Use this property to specify URL parameters that are to be set when the story is opened.
- `receivers`
  Use this property to specify the users who should receive the notification.
- `title`
  Use this property to specify the title of the notification.

[»]

Notifications by Email

To send notifications from SAP Analytics Cloud by email, you must configure a user-defined email server in the settings of your SAP Analytics Cloud tenant. You can find this via the **System • Administration • Notifications** path.

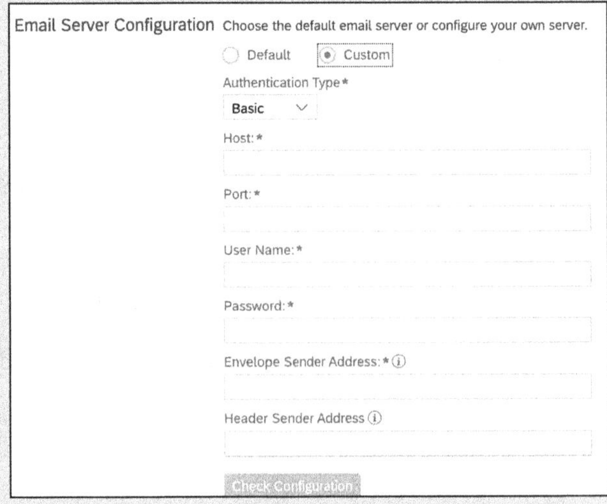

Figure 3.18 Configuring Email Server

To do this, you must enter the necessary information in the configuration of the email server (see Figure 3.18). You can then have the configuration checked by SAP Analytics Cloud and save the settings if successful.

Scheduling

Publications

The `Scheduling` API helps you to use the publication function from SAP Analytics Cloud in even greater detail. By default, you use `Scheduling` to send a story to end users in a specific format. You can use this API to check, among other things, whether the story is called up via a publication. The following methods are available to you in this API:

- **`isRunBySchulePublication`**
  This method checks whether the story is called via a publication.
- **`logMessage`**
  Use this method to log messages in the publication details.
- **`openSubscriptionDialog`**
  This method opens the dialog for creating a publication in your story. However, this function is not available on mobile devices.
- **`publish`**
  This method triggers publishing manually.

You can see an example of how to use these methods in the following code:

```
// Returns a true or false value in the console as to whether
// the application was called via a publication
 console.log(Scheduling.isRunBySchedulePublication());
```

[«]

Automated Notifications

By combining the `isRunByPublication` and `SendNotification` methods, you can send automated notifications.

URLParameter

You can use the `create` method of the `URLParameter` API to define a new URL parameter within your story. You can enter a name and content.

UserInfo

The `UserInfo` API returns information about the currently logged-in SAP Analytics Cloud user. You can access the display name and ID.

Timer

You can use the Timer API to control a *timer* defined in your application. You can check whether the timer is currently running and also start and stop it. An event can also be called up here when the timer has expired.

Here's an example of this:

```
// Starts the timer with a one-second delay
Timer_1.start(1);
// Logs in the console whether the timer is running or not
 console.log(Timer_1.isRunning());
```

Create Animations

If you want to create a visual highlight in your story, use the Timer API to run your key figures through the story. You can find an example in the content network in the Analytics Designer examples.

Widget

The methods of the Widget API are passed on to the widgets and can therefore be used for dropdown lists, filter lines, and other widgets. Among other things, you can find out or set the CSS class of a widget, show and hide a widget, or check whether it's currently visible.

These methods are also inherited by the subclasses, and you can use them in the APIs for the various widgets. For example, you can use the following command to hide Chart 1 from your story:

```
Chart_1.setVisible(false);
```

Generic Widget

You can use the Generic Widget API to obtain all widgets of a specific type, such as a chart or a table. Optionally, you can also enter a search term here. In this case, only the widgets that contain this term in their name are returned.

Case Sensitive

The search term is case sensitive, which means for example that a search query with the keywords *Sales Details* would not display any widgets with the name *sales details*.

The example in Listing 3.2 shows you how you can use this API.

```
// Returns an array of all chart widgets
var charts = Application.getWidgets({ type: WidgetType.Chart });
// Returns an array of all charts whose
// names contain the text "Sale"
var charts = Application.getWidgets({ type: WidgetType.Chart,
searchPattern: "Sale"});
```

Listing 3.2 Example for Application of Generic Widget API

3.2.2 Container

APIs for the various containers

In addition to the APIs of the **Standard** type library, there are various container objects in the analytical applications. These are usually used to group individual objects or keep them together. The methods of the container APIs are often available for different container widgets, and we therefore don't explain them for each widget below.

Flow Panel

One type of container object is the *flow panel*, whose functions relate to the flow layout panel widget, which is the object described in Section 3.1.4. You can show or hide a busy indicator by using the `hideBusyIndicator` and `showBusyIndicator` methods, or you can move widgets by using the `moveWidget` method.

You can see how to use these methods in your story in the example in Listing 3.3.

```
// Displays a busy indicator, moves Chart_1, and then hides the busy
indicator
//again
FlowLayoutPanel_1.showBusyIndicator("Please wait");
FlowLayoutPanel_1.moveWidget(Chart_1);
FlowLayoutPanel_1.hideBusyIndicator();
```

Listing 3.3 Example for Application of Flow Panel API

This example shows how you can use the flow panel. We've used this code in the `onInitialization` event. Here, a busy indicator with the message `"Please wait"` is displayed at the start of the initialization of the application, and `Chart 1` is then moved from the area outside the flow panel into the flow panel. Finally, the busy indicator is hidden again.

PageBook

With the PageBook API, you can access the book page widget. In addition to the APIs that you're already familiar with from other widgets (e.g. MoveWidget, the busy indicator), you can use your own methods here. For example, you can read out the currently selected book page or set the selected key.

The following example shows you how you can implement this:

```
// Writes the page to a variable and outputs it to the
// console
var Page = PageBook_1.getPage("Page_1");
console.log(Page);
```

Panel

The methods that you know from the Widget API are available for the panel widget, and you can use these methods to do things like moving an area or displaying a busy indicator.

Popup

In addition to the inherited methods from the widget, the API for the popup widget offers a number of custom methods to control the widget more precisely. For example, you can open and close the popup, and you can also read and set the title and control the buttons in the footer individually and show and hide them, as well as activate and deactivate them.

The following methods are available to you in the popup widget:

- **close**
  This method closes the popup.
- **getTitle or setTiltle**
  The first of these methods returns the title of the popup, and the second sets it.
- **isButtonEnabled or setButtonEnabled**
  The first of these methods returns whether a specific button in the footer of the popup is enabled, and the second enables or disables the button.
- **isButtonVisible** or **setButtonVisible**
  The first of these methods returns whether a specific button is visible in the footer of the popup, and the second makes it visible or hidden.
- **open**
  This method opens the popup.

You can open a popup with the following code, for example:

```
// Opens the desired popup
 Popup_1.open();
```

TabStrip/Tab

The TabStrip API for the tab bar and the Tab API for the individual tabs also inherit functions from the widget. However, additional methods are available to read or set the selected key, read and set the text, or read the selected tab.

This API also contains an onSelect event that is executed when the user selects a tab.

The following methods, among others, are available here:

- **MoveWidget**
  This method moves a widget to the defined tab of the tab strip.
- **getSelectedKey** or **setSelectedKey**
  The first of these methods returns the key of the selected tab, and the second sets it.
- **getTab**
  This method returns a tab of the tab strip.

3.2.3 Chart

Change diagrams with APIs

Another type library is the **Chart**. A wide variety of APIs are available for the charts you use in your story, and as these are available for all charts, we'll explain them in more detail in the following sections.

Chart

The Chart API contains the basic methods that are available for charts. The two most important events here are the onSelect event, which is called when the end user clicks on the chart with the mouse, and the onResultChanged event, which is called when the result of the chart changes.

There are also a number of methods available that you can use to customize the diagrams in your story or read out information. For example, you can add dimensions to your story or remove them, read out the data source of the chart, or activate and deactivate the chart. You can also open a new story, show or hide the context menu, and influence the quick actions menu.

The following methods, among others, are available here:

- **addDimension, getDimensions, or removeDimension**
  Use these methods to add or remove dimensions on an axis in your diagram or to read out which dimensions are on this axis. This is also possible with the Member and Selections types as well as with other types.
- **getDataChangeInsights**
  This method returns the data change insights of the diagram.

- `getDataSource`
  This method returns the data source of the diagram.
- `openInNewStory`
  This method creates a new story with the diagram.
- `setEnabled` or `isEnabled`
  The first of these methods activates interaction with the diagram, and the second deactivates it. You can also check whether the diagram is active or inactive.
- `setQuickActionsVisibility`
  The `setQuickActionsVisibility` method shows the quick actions as they are defined in the `ChartQuickActionsVisibility` component.
- `setContextMenuVisible`
  Use this method to set whether the **More actions** menu or the right-click menu should be displayed.

For example, you can use the `Chart` API to read the dimension from your diagram:

```
// Reads the dimensions of the chart from the trellis
 Chart_1.getDimensions(Feed.Trellis);
```

ChartNumberFormat

The `ChartNumberFormat` API allows you to adjust the number format of your key figures within the story. This is particularly useful for scaling or adjusting decimal numbers.

ChartQuickActionsVisibility

The `ChartQuickActionsVisibility` API allows you to customize the quick menu of your story. This allows you to decide in which individual situations which menu options will be available, and you don't have to define this permanently via the designer for the chart.

Feed

You can use the `Feed` API to customize the various feeds of your chart. You need this option, for example, if you want to add another key figure within your story. You must specify where this should be displayed.

3.2.4 Datasource

Access to the data source with APIs

To access your data source, you can use various APIs from the **Datasource** type library. These APIs allow you to access your data source and set filters or query details about your data, as we'll discuss in the following sections.

DataExplorer

With the `DataExplorer` API, you can close and open the data explorer, add dimensions to it, adjust the sorting of the dimensions, and show and hide additional dimensions.

DataSource

The `DataSource` API allows you to read information from your data source. This is particularly helpful for logging information in the console and thus checking the results and changes when executing the story. You can use this API to read out various types of information such as dimension filters, dimension values, dimensions, key figures, information about your data source, and much more. You can both read and set this information in this API.

[«]

Access Options

If you use the `DataSource` API, you can access both the ID and the description of the specific data source.

You can also check here whether the pause mode is set, and you can deactivate or activate it.

Another helpful method is to refresh the data or open the prompt dialog, which you can do with the following command:

```
// Refreshes the data of the data source of Chart 1
Chart_1.getDataSource().refreshData();
```

DataSourceInfo

You can use the `DataSourceInfo` API to obtain various information from the data source, such as information on the model used.

If you're using a business warehouse as a data source, you can retrieve further information such as the name or description of the source or the last modification date.

PauseMode

With the `PauseMode` API, you can adjust the pause mode setting in your story. The **Auto**, **On**, and **Off** options are available here, as in the right-side panel.

3.2.5 Visualization Controls

API access to additional widgets

In addition to the APIs for the charts, APIs are available for the visualization controls. There are different events that are shared by several visualization controls, so we explain them in the introduction to this section.

The `onClick` event is called when the user clicks on the visualization control, and the `onLongPress` event is called when a user presses and holds the visualization control. These two events are shared by the `Image`, `Shape`, and `Text` (only `onClick`) visualization controls.

Image

In addition to the two events already mentioned, the `Image` API has two methods. You can use these methods to set a desired image in your image widget or store a hyperlink that can be opened by clicking on the image.

RssReader

With the `RssReader` API, you can control your RSS reader. This gives you the option of adding new feeds or removing existing feeds. You can also use this API to read all feeds or the currently selected feed.

Shape

In addition to the two events already mentioned, you have the option of setting a hyperlink to the *shape* widget, which opens the desired URL with a click. You can also apply a style to the shape widget—the so-called *shape style*—which determines the fill and line color.

Text/TextStyle

You can use the `Text` API to replace existing text or read out the plain text without formatting, and you can use the `TextStyle` API to adjust the background color and the color of the font.

WebPage

With the `WebPage` API, you have the option of setting an address to which the web page should navigate. In addition, you can send a message to the web page using the `post` method.

3.2.6 Advanced Controls

APIs for geo maps and R visualizations

Using the APIs of the **Advanced Controls** type library, you can access geo maps and R visualizations in your story, as we'll cover in the following sections.

GeoMap

The `GeoMap` APIs extend the `Widget` API. There are a total of three different APIs for the *GeoMap* widget:

- With the `GeoMap` API, you have the option of retrieving the layers of your geo map or showing and hiding the `GeoMap Quick` actions.
- With the `GeoMapLayer` API, you can check whether a specific layer is currently visible and hide or show it.
- Use the `GeoMapQuickActionsVisibility` API to customize the quick menu of the geo map and thus deactivate the function to switch to full screen, for example.

RVisualization

With this API, you have the option of accessing your R visualization widget. In addition to the already familiar `onResultChanged` event, you have various methods to choose from.

You can read out specific R visualization elements such as the data frame or the environment values, and you can also read the contained input parameters, the status, and messages from the console of the R visualization. Furthermore, you have access to already known methods for setting the context menu, quick action menus, and hyperlinks.

3.2.7 Input Controls

APIs for input controls

Various APIs are also available for the different input controls that you can use to influence your story, as we'll discuss in the following sections. This gives you the option of accessing widgets with which you can process input.

Button

The `Button` API provides you with four methods to choose from in addition to the two `onClick` and `onLongPress` events: `getText`, `getToolTip`, `setText`, and `setTooltip`. You can use these methods to read or set the text of the button or tool tip.

CheckboxGroup, RadioButtonGroup, Dropdown, and ListBox

As the four `CheckboxGroup`, `RadioButtonGroup`, `Dropdown`, and `ListBox` APIs offer the same APIs for your widgets, we've combined them here.

With these APIs, you can add items to your widget or remove them, read the selected key or text, or set the selected key. With the list box, it's also possible to read or set multiple elements.

InputField and TextArea

The APIs for the *InputField* and *TextArea* widgets are also the same, which is why we've combined them here. You can use these APIs to read or set the current value or adjust the style. You can also check whether your widget is activated or changeable and set this accordingly.

Slider and RangeSlider

With the APIs for the Slider and RangeSlider widgets, you can read and set the minimum and maximum values of simple and range sliders, and you can also read and set the range or the value. The OnChange event is also available to you.

Switch

The API for the switch widget allows you to activate and deactivate your switch and also check whether it's active. You can also check whether the switch status is on or off and set it accordingly.

FilterLine

The API for the *FilterLine* widget lets you replace the existing model with a new model.

3.2.8 Table

The APIs in the **Table** type library offer you many different options for customizing tables. One very comprehensive method is the method for opening and closing the *navigation panel* (see Figure 3.19).

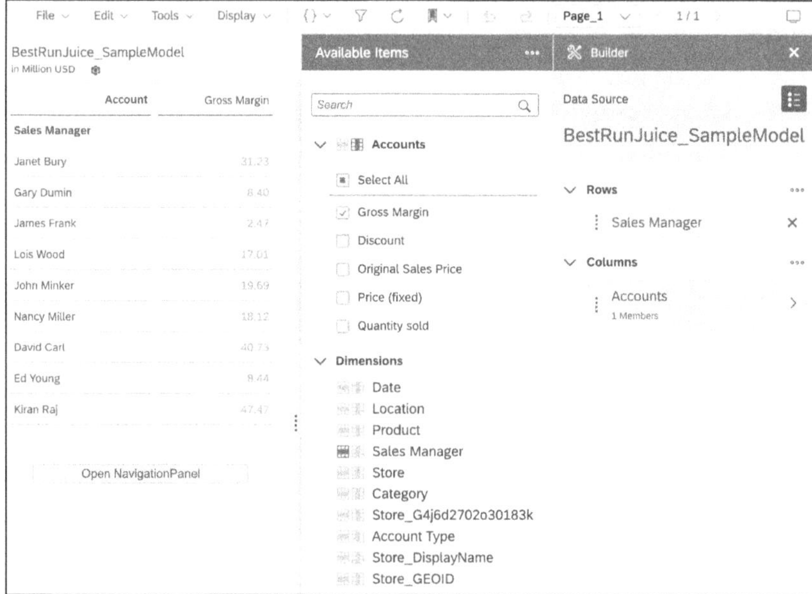

Figure 3.19 Opening Navigation Panel

This panel lets you or your end users add or remove rows or columns to or from your table without having to program these actions manually. There are also methods for adding dimensions to your rows or columns, and you can read out and replace the data source of your table or check and adjust the zero suppression. You can also open your widget with additional methods in a new story and adjust the ranking, sorting, or context menu. The `onResultchanged` and `onSelect` events are also available.

3.2.9 Export

Export APIs

Various APIs for exporting data from your applications in various formats are available in the **Exportcsv** and **Exportpdf** type libraries.

ExportCSV and ExportExcel

With the `ExportCSV` and `ExportExcel` APIs, your data can be exported in the form of a CSV or Excel file. You have the option of customizing the file name or the scope (only for CSV files) to be included in your file.

ExportPDF

With the `ExportPDF` API, you can export your story to a PDF document. In addition to the file name, you can change the header and footer of your file, as well as the size of the page. You can also add widgets or popups and perform the export in the background.

3.2.10 Bookmark

Develop individual bookmark logic

To save your filter settings in your story and call them up again, you can use the APIs of the **Bookmark** type library.

The most important API here is the `BookmarkSet` API. You can use this to assign, delete, or save bookmarks in your story. You can also use methods to display all bookmarks or only share the bookmarks in the story, and it's also possible to read out the version. You can also check whether the current status of the story is already saved in the bookmark or open the dialog for sharing bookmarks.

For example, you can save the following information in a bookmark:

- Width and height of all widgets
- Visibility of a widget
- Dimensions and key figures that have been added to a widget
- Selected values in a dropdown list
- Maximum and minimum values and the current value of a slider or range slider

3.3 Developing a Story with Scripting for Reporting

In Section 3.1 and Section 3.2, we showed you which widgets and APIs are available to you for developing a story in SAP Analytics Cloud. We would now like to use these elements to show you how you can develop your own story. To do this, we'll first show you which questions you should ask yourself before developing a story so that you can create a good story structure right from the start. Of course, new requirements can always lead to changes during development, so we'll show you the best way to build your story, and we'll take a closer look at sample applications from SAP.

3.3.1 What Questions Do I Ask Myself Before Development?

Questions before development

Before you start developing a new application, you should think about what the vision of your story should look like. This will save you unnecessary changes and additional work in the development process.

First of all, you should think about how your story should look. If you want to have a header and footer, you should include them from the start so that the entire layout of your story doesn't have to change later. You should also include an area between your header and footer for the main area so that you can arrange the elements better there.

For example, if you insert a header and footer and don't include an area for the space in between, the size of an inserted widget with a height of 50% will be half the size of the entire story. However, if you use an area for this space, then the size of an inserted widget with a height of 50% will be half the size of the area.

Your story should also most likely include some functions, such as a filter function. Here, too, you should start by thinking about the form in which you want to enable this function (e.g., via a filter line, via an `onSelect` event of a button click). You should also plan enough space for this.

Various end devices

Other important issues relate to which end devices you want to have access to your story. For example, you need to take several things into account for access via a smartphone, as there's significantly less space available on a smartphone display and certain elements will be hidden above a certain display size. However, if you know that your story will only be accessed on large screens, you can save yourself the effort of this development work.

As with any dashboard, you should also already have an idea of what data you want to display and how. If you use a lot of key figures, it's certainly helpful to include a key performance indicator area, and if you work with many charts, you should also plan enough space for them.

These are just a few examples of questions you should ask yourself and thoughts you should consider before developing your story. Of course, they may look different in your company due to certain guidelines, or you may already have the answers.

3.3.2 How Do I Build a Story?

Structure of a story

Let's first take a look at how you can implement the points from Section 3.3.1 in practice. First, you should create a new story as described in Chapter 2, Section 2.6. Then, in this section, we'll cover how to configure the basic settings for the story and walk you through two demo applications.

Story Settings

To open the basic settings for a story, click on the white part of the graphics area and open the **Format** area on the right-hand side of the page. Here, you can change the layout of the graphics area. In addition to classic settings such as background color and grid, you can set the size of the graphics area (see Figure 3.20).

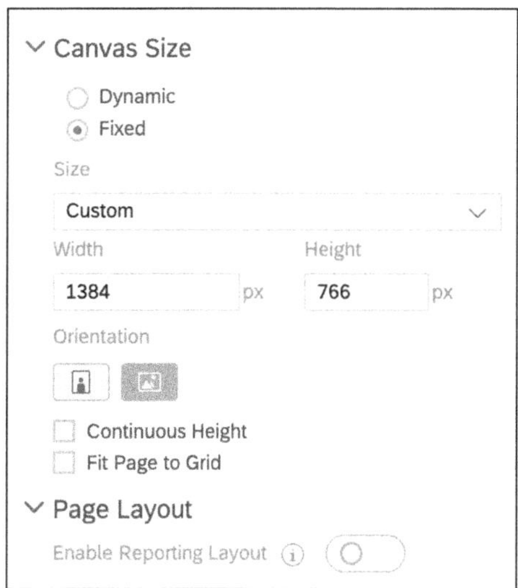

Figure 3.20 Settings of Fixed Graphics Area

You have the following size options:

- **Dynamic**
  The **Dynamic** option is very similar to the responsive story. Here, you can decide whether the layout of the content should be centered or adapted to the device. If you opt for the version adapted to the device, a device selection bar will be displayed in the graphics area (see Figure 3.21).

- **Fixed**
  The **Fixed** option is comparable to the graphic area variant from the story. Here, you can determine the size of the graphic area and set the alignment of the device.

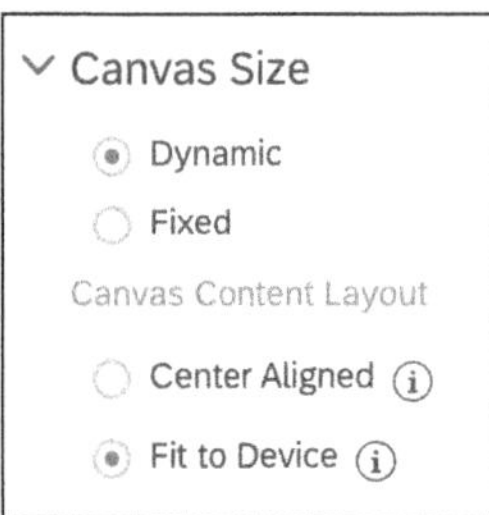

Figure 3.21 Settings of Dynamic Graphics Area

You can define CSS classes in the application settings. You can also define the automatic loading indicator and the text, and you can set the delay after which the indicator should be displayed (see Figure 3.22). If you want to use a PDF export in your story, you can choose here between the automatic option and the option via API.

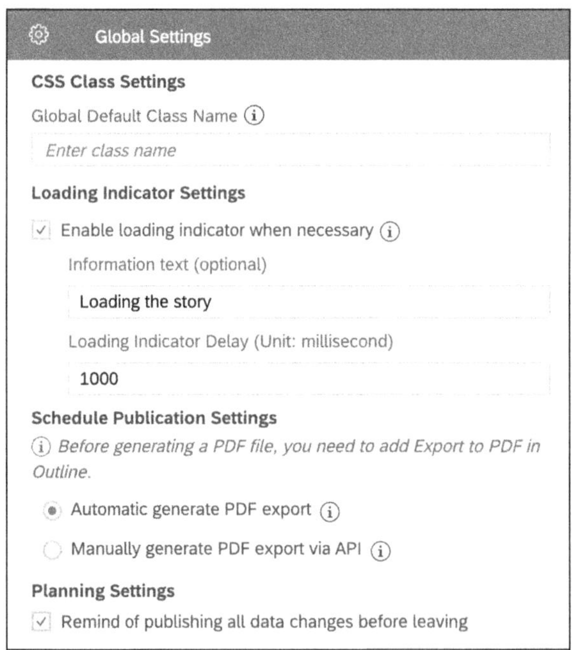

Figure 3.22 Application Settings of Story

Basic structure of a story

In the second step, we build a basic framework in which we define our areas and assign the desired sizes to them.

In our example, we have a header with 0% distance to the top and 95% distance to the bottom (see Figure 3.23). The spacing to the left and right is 0%, and the footer also extends across the entire width and has a distance of 95% to the top. Between the header and footer, we've inserted a main area that extends across the entire width of the screen and has a gap of 5% at the top and bottom of the screen. This main area also contains a filter area and a display area.

The filter area has a distance of 75% to the right and 5 % to the top and bottom of the main area (not to the top and bottom of the screen—this is where the different area elements come from). The display area takes up the remaining size. It has a distance of 25% to the left and no other distance.

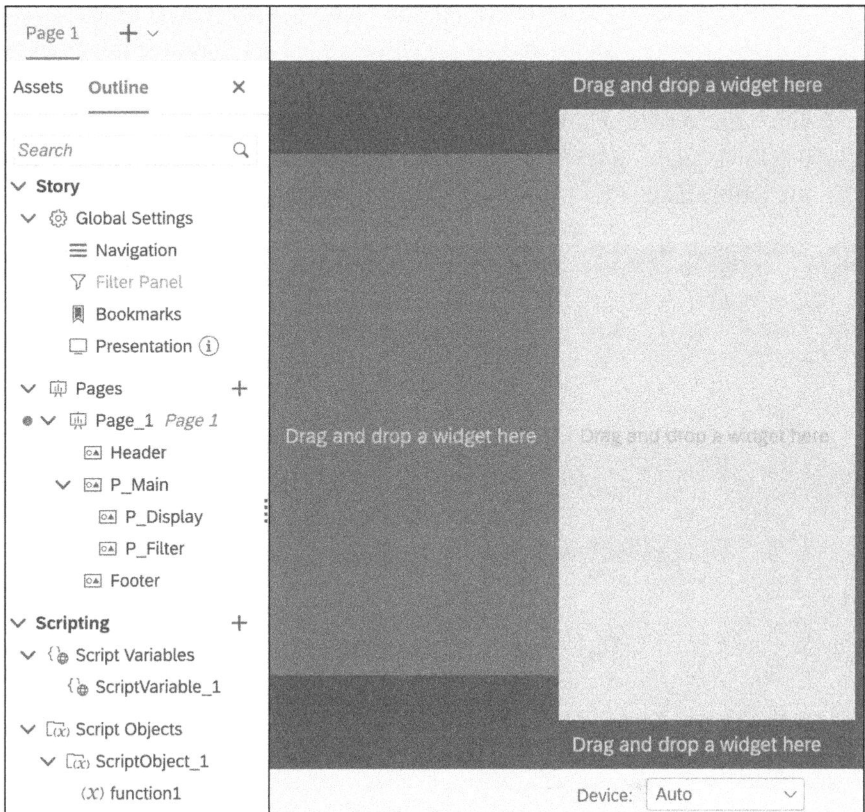

Figure 3.23 Example of Basic Structure of Story

You can now insert your widgets into this basic framework and use the various APIs.

To give you further impressions of what a story can look like, we would now like to take a closer look at two "old" applications from the content network.

Demo Application: Best-Run Sales

Structure of a demo application

The first demo application, **Best Run Sales,** provides an overview of various sales figures. We would like to show you what the application looks like in the finished version and which functions are available before we look at the technical details.

While the application is being called up, a loading indicator is displayed in the application display area, which corresponds to an image in the application. After the successful loading process, you'll land in the **Overview** area, and the loading indicator will disappear. In this area, you'll find several widgets that are not further organized. However, if you're developing your own applications, we would advise you to always create an outline if possible.

At the top of the application, you'll find a logo and the name of the application (see Figure 3.24). Next to it, you'll see the selected time period, which you can customize in the adjacent dropdown list. On the right-hand side, you'll find a text field with the current user and the current date. You can use the images on the left-hand side to switch between the **Overview** and **Analytics** areas.

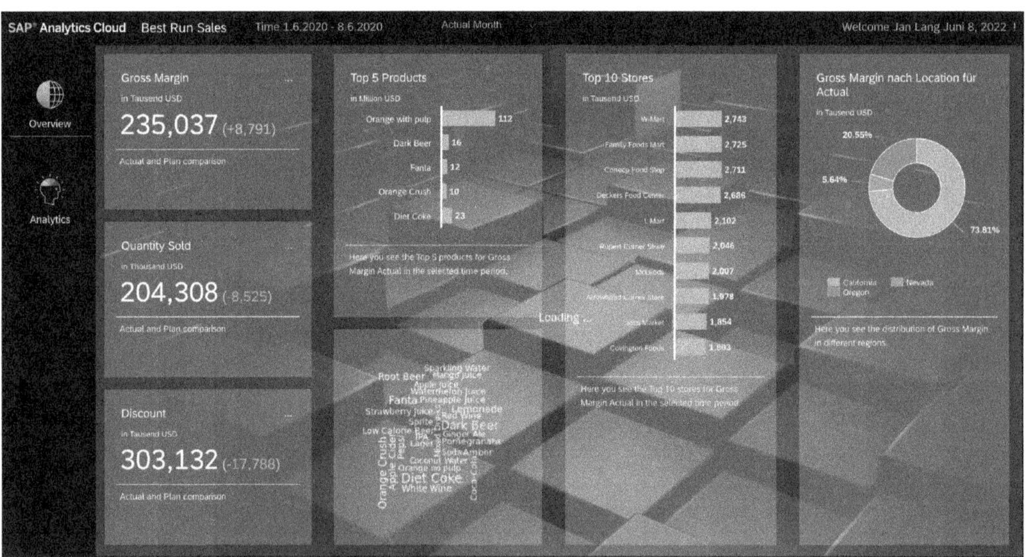

Figure 3.24 Best-Run Sales Demo Application

If you click on the logo in the **Analytics** area, you'll see the view in Figure 3.25. Here, you'll see a table by default, but you can use the icon bar on the right-hand side to customize the display to make it a chart. You can also use the **Show**, **by**, and **as** dropdown lists and a filter bar to customize the data displayed. As you can see, the header in the **Analytics** view is identical to the header in the **Overview**.

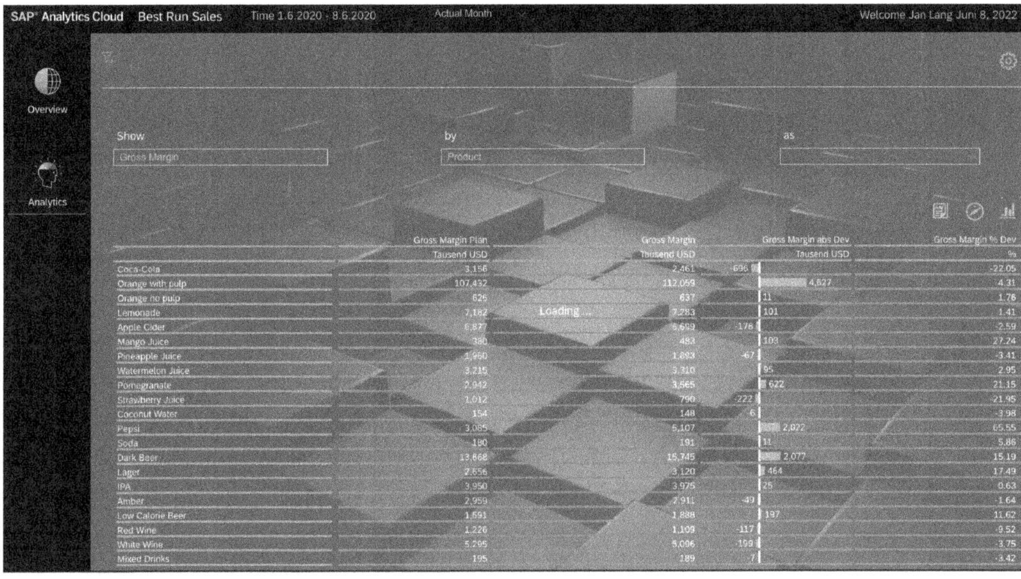

Figure 3.25 Analytics Section of Best-Run Sales Demo Application

Now that you've seen the interface of the finished application, let's take a look at the technical structure of the application. We'll start with the `onInitialization` event, which contains a console log that records the start and end of the event. In between, the key figures for the charts are set, among other things. This is done using various script objects so that they can be modified at a central location. After these script objects have been run through, the loading screen is hidden, and the **Overview** area is displayed via another script object (see Listing 3.4).

onInitialization event

```
console.log("onInitialization - begin");
// Set Measure for Chart.
Util_Layout.setMeasures();
// Set Chart or Table visible.
A_DropDown_Dimensions.setSelectedKey(CurrentDimension);
// Set Hierachy.
Util_Layout.setHierarchies();
// Set Time.
Util_Layout.setTime();
// Set URL for SmartDiscoverx.
A_CurrentSelection.applyText(' ');
Loading.setVisible(false);
Util_Layout.showOverview();
console.log('onInitialization - end');
```

Listing 3.4 Code of onInitialization Event

As examples of the different script methods used in this event, let's take a look at the `Util_Layout.setMeasures` and `Util_Layout.showOverview` script functions. Let's also start with the `setMeasures` object.

Settings for the popup

This mainly concerns the **Analytics** page. You use the object to read out which key figure was selected in the checkbox group or in the option field for the chart and the table, which were built into a popup. You open the corresponding popup by clicking the gear icon. The code in the `onClick` event is as follows:

```
Popup_Settings.open();
```

The selected key figures are written to variables and then set or removed in the chart and table. These are also logged in the console log. These script functions have been created under the `Util_Layout` script object with the `void` return type (see Figure 3.26).

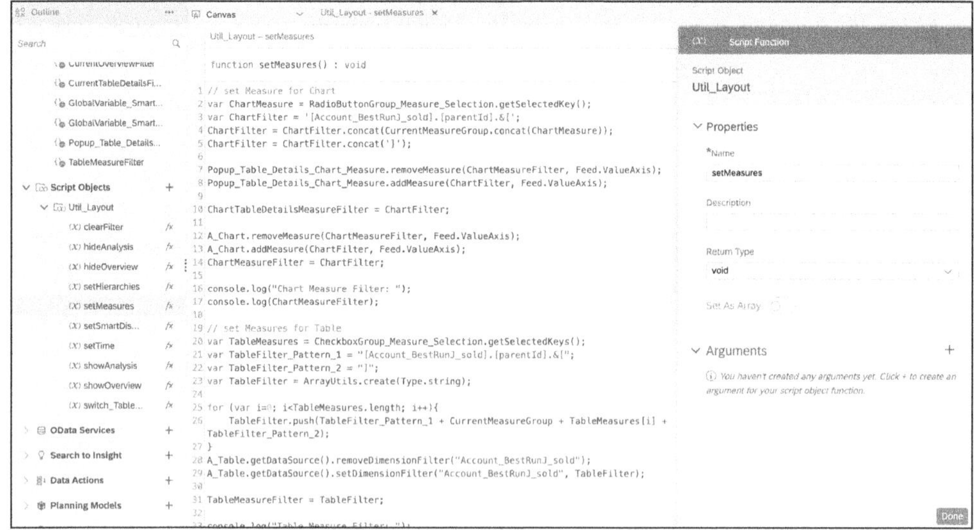

Figure 3.26 setMeasures Script Function

In the code example in Listing 3.5, in addition to temporary variables for this event, other data is made available in the script variables. You can assign default values to them and overwrite them via events and methods within the application (see Figure 3.27).

```
// Set Measure for Chart.
var ChartMeasure = RadioButtonGroup_Measure_
Selection.getSelectedKey();
var ChartFilter = '[Account_BestRunJ_sold].[parentId].&[';
ChartFilter =
ChartFilter.concat(CurrentMeasureGroup.concat(ChartMeasure));
ChartFilter = ChartFilter.concat(']');
```

```
Popup_Table_Details_Chart_Measure.removeMeasure(ChartMeasureFilter,
Feed.ValueAxis);
Popup_Table_Details_Chart_Measure.addMeasure(ChartFilter,
Feed.ValueAxis);
ChartTableDetailsMeasureFilter = ChartFilter;
A_Chart.removeMeasure(ChartMeasureFilter, Feed.ValueAxis);
A_Chart.addMeasure(ChartFilter, Feed.ValueAxis);
ChartMeasureFilter = ChartFilter;
console.log("Chart Measure Filter: ");
console.log(ChartMeasureFilter);
// Set Measures for Table.
var TableMeasures = CheckboxGroup_Measure_Selection.getSelectedKeys();
var TableFilter_Pattern_1 = "[Account_BestRunJ_sold].[parentId].&[";
var TableFilter_Pattern_2 = "]";
var TableFilter = ArrayUtils.create(Type.string);
for (var i=0; i<TableMeasures.length; i++){
TableFilter.push(TableFilter_Pattern_1 + CurrentMeasureGroup +
TableMeasures[i] + TableFilter_Pattern_2);
}
A_Table.getDataSource().removeDimensionFilter("Account_BestRunJ_
sold");
A_Table.getDataSource().setDimensionFilter("Account_BestRunJ_sold",
TableFilter);
TableMeasureFilter = TableFilter;
console.log("Table Measure Filter: ");
console.log(TableMeasureFilter);
```

Listing 3.5 Code of setMeasures Function

Figure 3.27 ChartTableDetailsMeasureFilter Script Variable

Show and hide areas

Next, let's take a look at the showOverview function, which ensures that all objects that don't belong to the **Overview** area are hidden and, in return, shows all associated objects. A pardon for the Analytics area was also created in the function, and each of the following script objects was placed on the onClick event of the selected icon (see Listing 3.6).

```
H_Menu_Overview.setVisible(false);
H_Menu_Overview_Clicked.setVisible(true);
O_Discount_Bg.setVisible(true);
O_Discount_Chart.setVisible(true);
O_Discount_Details.setVisible(true);
O_Discount_Text.setVisible(true);
O_GrossMargin_Bg.setVisible(true);
O_GrossMargin_Chart.setVisible(true);
O_GrossMargin_Details.setVisible(true);
O_GrossMargin_Text.setVisible(true);
O_Location_Bg.setVisible(true);
O_Location_Chart.setVisible(true);
O_Location_Text.
setVisible(true);
O_Location_Border.setVisible(true);
O_Product_Bg.setVisible(true);
O_Product_Chart.setVisible(true);
O_Product_Text.setVisible(true);
O_Product_Cloud_Bg.setVisible(true);
O_Product_Cloud_RVis.setVisible(true);
O_Quantity_Bg.
setVisible(true);
O_Quantity_Chart.setVisible(true);
O_Quantity_Details.setVisible(true);
O_Quantity_Text.setVisible(true);
O_Stores_Bg.setVisible(true);
O_Stores_Chart.setVisible(true);
O_Stores_Text.setVisible(true);
```

Listing 3.6 Code for ShowOverview Function

Linked analysis

The onSelect events of the individual charts, such as the O_Stores_Chart chart, are also stored with code. Here, the selected value or values are stored in a variable, which is passed on to the other charts on this page as a filter. This gives the user the option of filtering all other charts according to the location (Oregon, for example) by selecting the location within the charts. You can see the code for the onSelect event in Listing 3.7.

```
var sel = O_Stores_Chart.getSelections();
console.log( ['Chart Selection: ', sel, CurrentOverviewFilter ]);
if (sel.length > 0) {
```

```
var selection = sel[0];
for (var dimensionId in selection) {
if (dimensionId === "Store_3z2g5g06m4") {
var memberId = selection[dimensionId];
console.log( ['Chart Selection Dimension: ', dimensionId]);
console.log( ['Chart Selection Member: ', memberId]);
if (CurrentOverviewFilter !== '') {
console.log( ['Chart Selection: ', "Clear Filter",
CurrentOverviewFilter ]);
O_GrossMargin_Chart.
getDataSource().removeDimensionFilter( CurrentOverviewFilter
);
O_Discount_Chart.getDataSource().removeDimensionFilter(
CurrentOverviewFilter);
O_Quantity_Chart.getDataSource().removeDimensionFilter(
CurrentOverviewFilter);
Chart_Discount_Details.getDataSource().removeDimensionFilter(
CurrentOverviewFilter
);
Chart_Gross_Margin_Details.getDataSource().removeDimensionFilter(
CurrentOverviewFilter);
Chart_Quantity_Sold_Details.getDataSource().removeDimensionFilter(
CurrentOverviewFilter);
CurrentOverviewFilter = '';
}
CurrentOverviewFilter = "Store_3z2g5g06m4";
O_GrossMargin_Chart.getDataSource().setDimensionFilter(
dimensionId, memberId)
;
O_Discount_Chart.getDataSource().setDimensionFilter( dimensionId
, memberId);
O_Quantity_Chart.getDataSource().setDimensionFilter(
dimensionId, memberId);
Chart_Discount_Details.getDataSource().setDimensionFilter(
dimensionId, memberId
);
Chart_Gross_Margin_Details.getDataSource().setDimensionFilter(
dimensionId, memberId);
Chart_Quantity_Sold_Details.getDataSource().setDimensionFilter(
dimensionId, memberId);
}
    }
 }
```

Listing 3.7 onSelect Event of Chart

In addition, various popups were created in this application. The standard header and footer were not used in these but were instead created using text fields and buttons. The popups mostly contain a chart and a button that closes the popup.

Popup for the settings

Only the popup for the settings is a little more complex, but this complexity is not apparent at first glance. Depending on whether the chart or the table is currently displayed in the application, the popup contains either an option field or a group of checkboxes. You can select values here, and if you then click the **Cancel button**, the popup is closed and nothing changes. However, if you click the **OK** button, the chosen selections are read out and set in the chart or table.

Here, it would certainly be a good idea to outsource the code to a script function and close the popup afterward. In the example, however, there's no script function except the code in Listing 3.8:

```
// Set Measure for Chart.
var ChartMeasure = RadioButtonGroup_Measure_Selection.getSelectedKey();
var ChartFilter = '[Account_BestRunJ_sold].[parentId].&[';
ChartFilter =
ChartFilter.concat(CurrentMeasureGroup.concat(ChartMeasure));
ChartFilter = ChartFilter.concat(']');
A_Chart.removeMeasure(ChartMeasureFilter, Feed.ValueAxis);
A_Chart.addMeasure(ChartFilter, Feed.ValueAxis);
// Set new Measure for Popup_Chart_Details_Chart_Details.
Popup_Chart_Details_Chart_Details.removeMeasure(ChartMeasureFilter,
Feed.ValueAxis);
Popup_Chart_Details_Chart_Details.addMeasure(ChartFilter,
Feed.ValueAxis);
ChartMeasureFilter = ChartFilter;
console.log(["Chart Measure Filter: ", ChartMeasureFilter]);
// Set Measures for Table.
var TableMeasures = CheckboxGroup_Measure_Selection.getSelectedKeys();
var TableFilter_Pattern_1 = "[Account_BestRunJ_sold].[parentId].&[";
var TableFilter_Pattern_2 = "]";
var TableFilter = ArrayUtils.create(Type.string);
for (var i=0; i<TableMeasures.length; i++){
TableFilter.push(TableFilter_Pattern_1 + CurrentMeasureGroup +
TableMeasures[i] + TableFilter_Pattern_2);
}
A_Table.getDataSource().removeDimensionFilter("Account_BestRunJ_sold");
A_Table.getDataSource().setDimensionFilter("Account_BestRunJ_sold",
TableFilter);
// Smart Discovery.
```

```
// Table_SmartDiscoveryReference.getDataSource().
// removeDimensionFilter("Account_BestRunJ_sold"); // Table_
SmartDiscoveryReference.
getDataSource().
// setDimensionFilter("Account_BestRunJ_sold", TableFilter);
TableMeasureFilter = TableFilter;
console.log(["Table Measure Filter: ", TableMeasureFilter]);
Popup_Settings.close();
```

Listing 3.8 Close Settings Popups

Demo Application: Reporting Dashboard

SAP Fiori look

The next application from the Content Network that we'll be presenting to you is the reporting dashboard. This application is based on the new SAP Fiori design guidelines from SAP.

[«]

> **Clearing Up Biases**
>
> There has always been a biased public perception that SAP applications are cumbersome to use and don't look particularly appealing. With *SAP Fiori*, SAP attempts to counteract this bias. SAP Fiori is both the guideline for how SAP wants to design the user experience and the name of the new SAP user interface.

The demo application has a header area in which you'll find a logo and the name of the application on the left-hand side. On the right-hand side, there's a search icon [search icon], an image, and another text field (see Figure 3.28). Below this header area is a tab bar with three different tabs, and the **Home** tab is particularly noticeable here. The user can modify this individually with tiles from the other two tabs, each of which contains different charts. So, let's first take a closer look at the header area and the **Home** tab.

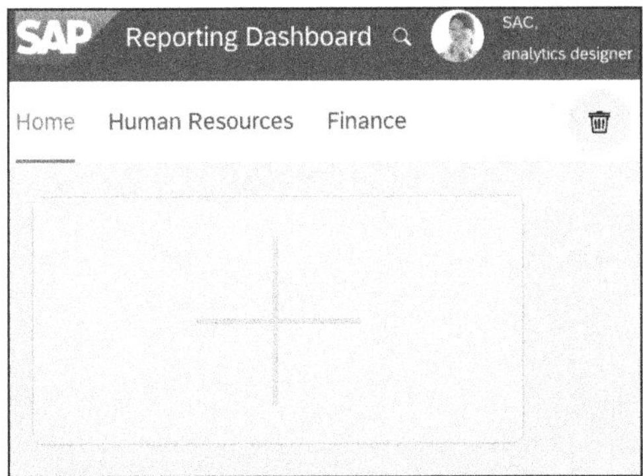

Figure 3.28 Home Area of Application

Let's start with the onInitialization event here. This event first checks whether the user has saved tiles in their personal home screen. These are saved in bookmarks, and a corresponding bookmark group has been created in the **Outline** area for this purpose (see Figure 3.29).

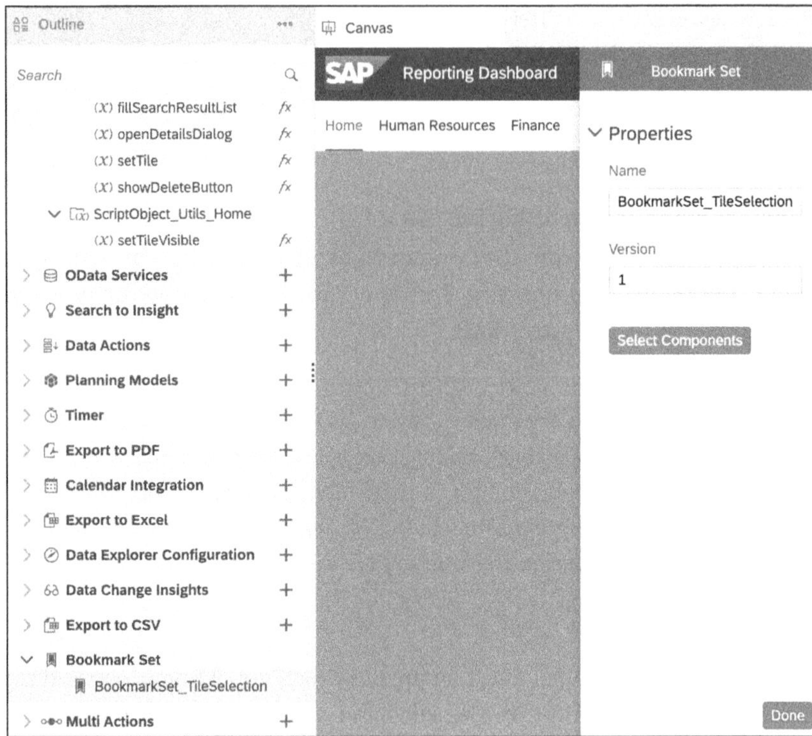

Figure 3.29 Bookmark Group in Application

First, the system checks whether bookmarks are available, and then, it runs through various script elements (see Listing 3.9).

```
// Is not bookmark, check for bookmark
debugger.
if (isBookmark === 0) {
var bookmarks = BookmarkSet_TileSelection.getAll();
if (bookmarks.length > 0) {
var bookmark = bookmarks[0];
BookmarkSet_TileSelection.apply(bookmark);
}
}
ScriptObject_Utils.defineTiles();
ScriptObject_Utils.fillAddTileList();
ScriptObject_Utils.fillSearchResultList();
```

```
ScriptObject_Utils_Home.setLayout();
ScriptObject_Utils_HR.setLayout();
ScriptObject_Utils_Finance.setLayout();
```

Listing 3.9 Open Bookmarks

Check screen size

The `ScriptObject_Utils_Finance.setLayout()` script object is particularly interesting here. It checks the size of the screen used to call up the application and adjusts the size of the widgets accordingly. The position of the selected tile is determined via another script object (see Listing 3.10).

```
var screenWidth = Application.getInnerWidth().value;
console.log(["setLayout HR : ", screenWidth]);
// 1 column
if (screenWidth <= 570) {
Tile_Employees.getLayout().setTop(20);
Tile_Employees.getLayout().setLeft(20);
Tile_Employees.getLayout().setHeight(150);
Tile_Entries.getLayout().setHeight(150);
Tile_Seperations.getLayout().setHeight(150);
ScriptObject_Utils_Home.setPosition(screenWidth, Tile_Employees,
Tile_Entries, 150, 300, 20);
ScriptObject_Utils_Home.setPosition(screenWidth, Tile_Entries, Tile_
Seperations, 150, 300, 20);
ScriptObject_Utils_Home.setPosition
(screenWidth, Tile_Seperations, Tile_Sickness, 150, 300, 20);
ScriptObject_Utils_Home.setPosition(screenWidth, Tile_Sickness, Tile_
Average, 150, 300, 20);
ScriptObject_Utils_Home.setPosition
(screenWidth, Tile_Average, Tile_Overtime, 150, 300, 20);
ScriptObject_Utils_Home.setPosition(screenWidth, Tile_Overtime, Tile_
Surplus_Percent, 150, 300, 20);
ScriptObject_Utils_Home.
setPosition(screenWidth, Tile_Surplus_Percent, Tile_Surplus_Absolute,
150, 300, 20);
ScriptObject_Utils_Home.
setPosition(screenWidth, Tile_Surplus_Absolute, Tile_Vacancy, 150,
300, 20);
ScriptObject_Utils_Home.setPosition(screenWidth, Tile_Vacancy, Tile_
Quota, 150, 300, 20);
// 2 column
} else if (screenWidth <= 812) {
Tile_Employees.getLayout().setTop(20);
Tile_Employees.getLayout().setLeft(20);
Tile_Employees.getLayout().setHeight(490);
```

```
Tile_Entries.getLayout().setTop(20);
    Tile_Entries.getLayout().setLeft(340);
Tile_Entries.getLayout().setHeight(490);
Tile_Seperations.getLayout().setTop(530);
Tile_Seperations.getLayout().setLeft(20);
Tile_Seperations.
getLayout().setHeight(490);
Tile_Sickness.getLayout().setTop(530);
Tile_Sickness.getLayout().setLeft(340);
Tile_Average.getLayout().setTop(700);
Tile_Average.
getLayout().setLeft(340);
Tile_Overtime.getLayout().setTop(870);
Tile_Overtime.getLayout().setLeft(340);
ScriptObject_Utils_HR.setPosition(screenWidth, Tile_Overtime, Tile_
Surplus_Percent, 150, 300, 20, 340);
ScriptObject_Utils_HR.setPosition(screenWidth, Tile_Surplus_Percent,
Tile_Surplus_Absolute, 150, 300, 20, 340);
ScriptObject_Utils_HR.setPosition(screenWidth, Tile_Surplus_Absolute,
Tile_Vacancy, 150, 300, 20, 340);
ScriptObject_Utils_HR.setPosition(screenWidth, Tile_Vacancy, Tile_
Quota, 150, 300, 20, 340);
} else {
Tile_Employees.getLayout().setTop(20);
Tile_Employees.getLayout().setLeft(20);
Tile_Employees.getLayout().setHeight(490);
Tile_Entries.getLayout
().setTop(20);
Tile_Entries.getLayout().setLeft(340);
Tile_Entries.getLayout().setHeight(490);
Tile_Seperations.getLayout().setTop(20);
Tile_Seperations.
getLayout().setLeft(660);
Tile_Seperations.getLayout().setHeight(490);
Tile_Sickness.getLayout().setTop(20);
Tile_Sickness.getLayout().setLeft(980)
;
    ScriptObject_Utils_HR.setPosition(screenWidth, Tile_Sickness,
Tile_Average, 150, 300, 20, 980);
ScriptObject_Utils_HR.setPosition
(screenWidth, Tile_Average, Tile_Overtime, 150, 300, 20, 980);
ScriptObject_Utils_HR.setPosition(screenWidth, Tile_Overtime, Tile_
Surplus_Percent, 150, 300, 20, 980);
ScriptObject_Utils_HR.setPosition(screenWidth, Tile_Surplus_Percent,
```

```
Tile_Surplus_Absolute, 150, 300, 20, 980);
ScriptObject_Utils_HR.setPosition(screenWidth, Tile_Surplus_Absolute,
Tile_Vacancy, 150, 300, 20, 980);
ScriptObject_Utils_HR.setPosition(screenWidth, Tile_Vacancy, Tile_
Quota, 150, 300, 20, 980);
}
```

Listing 3.10 Adapting Layouts to Screen Size

After loading the application, you'll see the header and the start area.

Perform searches in an application

Let's take a closer look at the search icon in the header. Click on it to open a popup (see Figure 3.30), which will contain an input field that you can use to enter a search term and the magnifying glass icon that you can use to start the search. Below this will be a group of checkboxes that you can use to select a tile.

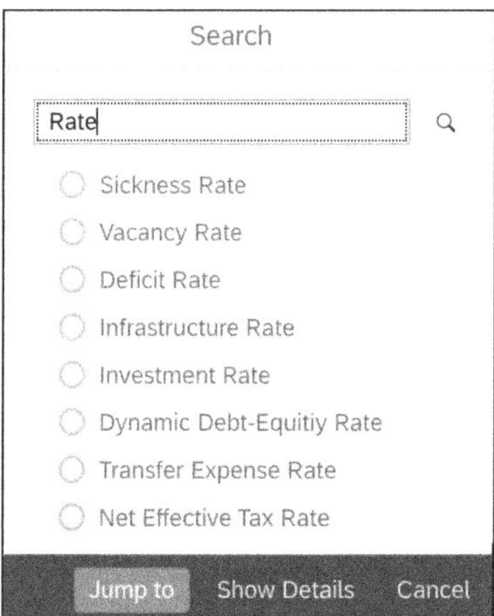

Figure 3.30 Popup for Searching for Tile

You can see the code for mapping the search logic in Listing 3.11.

```
var hit = false;
var number_hits = 0;
var searchstring = "";
searchstring = Popup_Search_Tile_Input.getValue();
Popup_Search_Tile_Result.setVisible(false)
;
// Slightly better handling of upper and lower case.
var abc = [["a", "A"],["b", "B"],["c", "C"],["d", "D"],["e", "E"],[
"f", "F"],["g", "G"],["h", "H"],["i", "I"],["j","J"],["k","K"],[
"l","L"],["m","M"],["n","N"],["o","O"],["p","P"],["q","Q"],[
"r","R"],["s","S"],["t","T"],["u","U"],["v","V"],["w","W"],[
```

```
"x","X"],["y","Y"],["z","Z"]];
var letter_large = "";
var letter_small = "";
var search_string_beginning_large = ""; //
var search_string_beginning_small = ""; //
// Search for the capital letter at the beginning.
 for (var i = 0; i < abc.length; i++) {
var letter = abc[i];
letter_large = letter[1];
letter_small = letter[0];
if (suchstring.substring(0,1) === letter_large) {
suchstring_start_large =.
suchstring.replace(suchstring.substring(0,1), letter_large);
suchstring_start_small =
suchstring.replace(suchstring.substring(0,1), buchstabe_klein);
} else if (suchstring.substring(0,1) === buchstabe_klein){
suchstring_anfang_gross =
suchstring.replace(suchstring.substring(0,1), buchstabe_gross);
suchstring_anfang_klein =
suchstring.replace(suchstring.substring(0,1), buchstabe_klein);
}
};
console.log(["suchstring_start_big",suchstring_start_big,
"suchstring_start_small",suchstring_start_small]);
// Initially remove all items.
Popup_Search_Tile_Result.removeAllItems();
for (i = 0; i < G_Search_Tiles.length; i++) {
var tile = G_Search_Tiles[i].split(";");
var tileKey = tile[0];
var tileDesc = tile[1];
if (tileDesc.indexOf(suchstring) >= 0 || tileDesc.indexOf(suchstring_
start_large) >= 0 || tileDesc.indexOf(suchstring_start_small) >= 0){
// Show the list box on the first hit and
create
a
// dummy entry. if (hit === false){
          treffer = true;
Popup_Search_Tile_Result.setVisible(true);
}
Popup_Search_Tile_Result.addItem(tileKey, tileDesc);
// Remember the number of hits. anzahl_treffer = anzahl_treffer + 1;
}
 };
```

Listing 3.11 Mapping Search Logic

In the footer of the popup, you can now decide whether you want to jump to this tile by clicking the **Jump to** button or whether you want to open the details of the popup for this tile by clicking the **Show details** button.

Add tiles

In the **Home** tab, you'll initially only see one tile with a large **+** sign. You can use this tile to add new tiles from the other two tabs, and you can also open a popup here (see Figure 3.31) in which there's a group of checkboxes that you can use to select a desired tile. Click **OK** to add the tiles to your home area and save them in a bookmark.

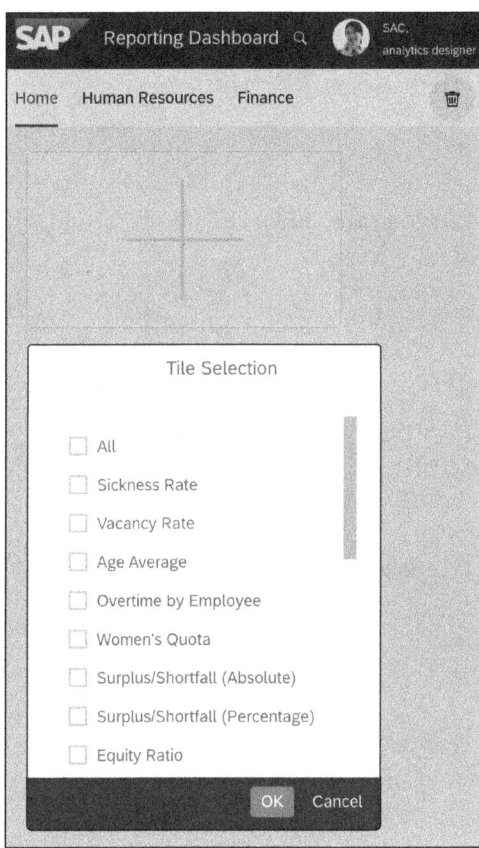

Figure 3.31 Add New Tile

This process is carried out as shown in Listing 3.12.

```
if (buttonId === "Btn_OK") {
var selections = Popup_Add_Tile_List.getSelectedKeys();
if (selections[0] === "ALL") {
selections = G_All_Tiles;
G_All_Tiles = EmptyArray;
}
```

```
      for (var i = 0; i < selections.length; i++) {
var sel = selections[i];
var j = Number.parseInt(G_All_Tiles.indexOf(sel).toFixed())+1;
console.log(["Indexof ", sel, j]);
G_All_Tiles = G_All_Tiles.slice(j);
switch (sel) {
case "T01":
ScriptObject_Utils.setTile(O_Tile_Sickness, Add_Tile);
Popup_Add_Tile_List.removeItem(sel);
break; case "
T02":
ScriptObject_Utils.setTile(O_Tile_Vacancy, Add_Tile);
Popup_Add_Tile_List.removeItem(sel);
break;
case "T03":
ScriptObject_Utils.setTile(O_Tile_Average, Add_Tile);
Popup_Add_Tile_List.removeItem(sel); break
; case "
T04":
ScriptObject_Utils
.
setTile(O_Tile_Overtime, Add_Tile);
Popup_Add_Tile_List.removeItem(sel);
break; case "
T05":
ScriptObject_Utils.setTile(O_Tile_Quota, Add_Tile);
Popup_Add_Tile_List.
removeItem(sel);
break;
case "T06":
ScriptObject_Utils.setTile(O_Tile_SurplusAbsolute, Add_Tile);
Popup_Add_Tile_List.removeItem(sel);
break;
case "T07":
ScriptObject_Utils.setTile(O_Tile_SurplusPercent, Add_Tile);
Popup_Add_Tile_List.removeItem(sel);
break;
case "T08":
ScriptObject_Utils.setTile(O_Tile_Equity_Ratio, Add_Tile);
Popup_Add_Tile_List.removeItem(sel); break
; case "
T09":
          ScriptObject_Utils.setTile(O_Tile_Expense, Add_Tile);
Popup_Add_Tile_List.removeItem(sel); break
```

```
; case "
T10":
ScriptObject_Utils.setTile(O_Tile_Expense_2, Add_Tile);
Popup_Add_Tile_List.removeItem(sel);
break; case "
T11":
ScriptObject_Utils.setTile(O_Tile_Deficit_Rate, Add_Tile);
Popup_Add_Tile_List.
removeItem(sel);
break;
case "T12":
ScriptObject_Utils.setTile(O_Tile_Infrastructure_Rate, Add_Tile);
Popup_Add_Tile_List.removeItem(sel);
break;
case "T13":
ScriptObject_Utils.setTile(O_Tile_Depreciation_Rate, Add_Tile);
Popup_Add_Tile_List.removeItem(sel); break
; case "
T14":
ScriptObject_Utils.
setTile(O_Tile_ThirdParty, Add_Tile);
Popup_Add_Tile_List.removeItem(sel);
break; case "
T15":
ScriptObject_Utils.setTile(O_Tile_Investment, Add_Tile);
Popup_Add_Tile_List.removeItem(sel);
break;
case "T16":
ScriptObject_Utils.setTile(O_Tile_Asset_Coverage, Add_Tile);
Popup_Add_Tile_List.
removeItem(sel);
break;
case "T17":
ScriptObject_Utils.setTile(O_Tile_Dynamic_Debt, Add_Tile);
Popup_Add_Tile_List.removeItem(sel);
break;
case "T18":
ScriptObject_Utils.setTile(O_Tile_Liquidity, Add_Tile);
Popup_Add_Tile_List.removeItem(sel); break
; case "
T19":
ScriptObject_Utils.setTile(O_Tile_Current_Lia, Add_Tile);
          Popup_Add_Tile_List.removeItem(sel);
break; case "
```

```
T20":
ScriptObject_Utils.setTile(O_Tile_Interest, Add_Tile);
Popup_Add_Tile_List.removeItem(sel);
break;
case "T21":
ScriptObject_Utils.
setTile(O_Tile_Labor, Add_Tile);
Popup_Add_Tile_List.removeItem(sel);
break; case "
T22":
ScriptObject_Utils.setTile(O_Tile_Material, Add_Tile);
Popup_Add_Tile_List.removeItem(sel);
break;
case "T23":
ScriptObject_Utils.setTile(O_Tile_Transfer, Add_Tile);
Popup_Add_Tile_List.removeItem(sel);
break; case "
T24":
ScriptObject_Utils.setTile(O_Tile_Net, Add_Tile);
Popup_Add_Tile_List.
removeItem(sel);
break;
}
if (G_Tile_Selection === "") {
G_Tile_Selection = sel;
} else {
G_Tile_Selection = G_Tile_Selection + ":" + sel;
}
console.log(["newTile" , sel, "G_Tile_Selection", G_Tile_Selection]);
}
}
console.log(["G_All_Tiles", G_All_Tiles]);
Popup_Add_Tile.close();
isBookmark = 1;
BookmarkSet_TileSelection.save("Reporting_Dashboard", false, true);
```

Listing 3.12 Add Tiles

Now that we've looked at the header area and the **Home** tab in detail, let's take a look at the two remaining tabs of the application: **Human Resources** and **Finance** (see Figure 3.32).

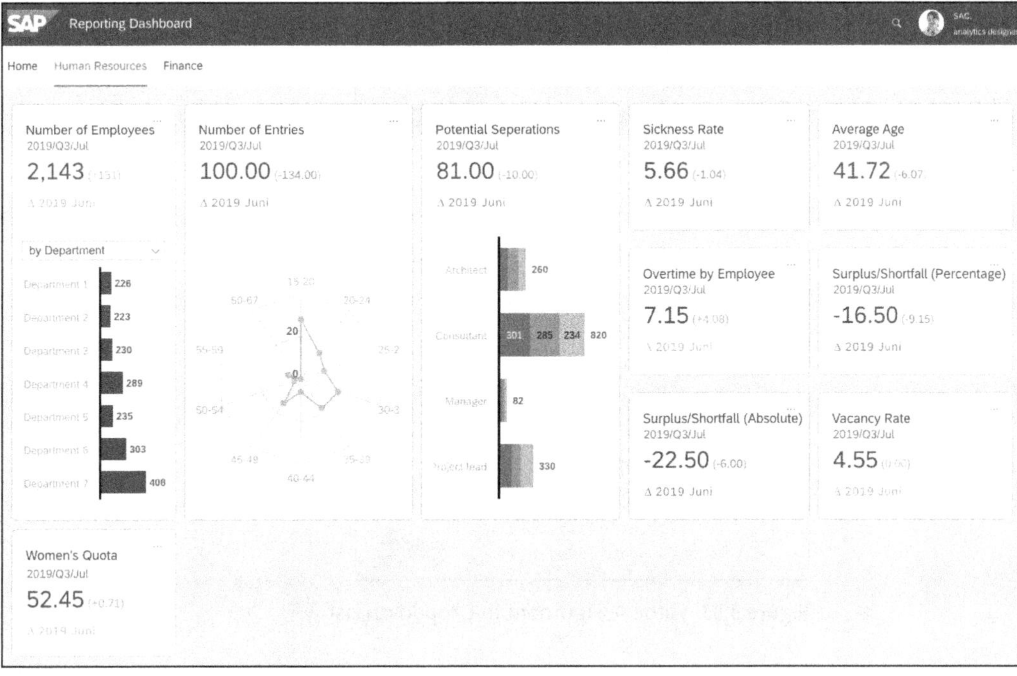

Figure 3.32 Further Tabs of Application

In these two tabs, you'll find various tiles, each of which contains information on different key figures. Some of these tiles are underpinned with additional charts, and in the tile for the number of employees, you can adjust the breakdown by using a dropdown list and breaking it down by another dimension. The current key figure is removed from the onSelect event, and the value selected in the dropdown list is read out and then added to the chart on the corresponding axis (see Listing 3.13).

```
Tile_Employees_Chart_Details.removeDimension(G_Tile_Nr_Employee_
CurrentDim,Feed.CategoryAxis);
G_Tile_Nr_Employee_CurrentDim = this.getSelectedKey();
Tile_Employees_Chart_Details.addDimension(G_Tile_Nr_Employee_
CurrentDim,Feed.CategoryAxis);
```

Listing 3.13 Remove Key Figure and Replace with Selection from Dropdown List

The values of the dropdown list were already permanently assigned during the development of the application and are not read from the data source (see Figure 3.33).

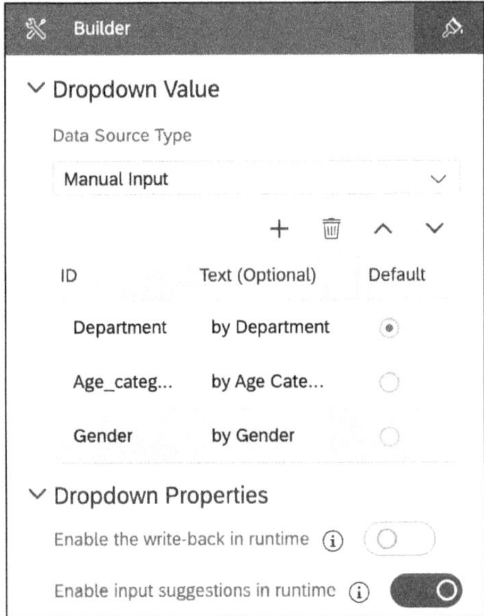

Figure 3.33 Value Assignment in Dropdown List

You also have the option of calling up a popup via the three dots in the top right-hand corner of each tile. Then, the code you entered in Listing 3.14 is executed.

```
console.log("Button OnClick");
G_DetailsFor = "Entries";
ScriptObject_Utils.openDetailsDialog(Popup_Details_Table_Entries,
"Number of Entries","[Key_70511202r1].[parentId].&[1002]");
```

Listing 3.14 Open Detail Popups

Dynamic opening of popups

In this onClick event, the popup is not called directly, but a script object is. The corresponding parameters for the tile are set in this object, and the popup is then opened (see Figure 3.34). This is particularly helpful if you want to call up the same popup with different numbers in several places (see Listing 3.15). In this way, you only have to create one popup and not a separate popup for each key figure. The initial effort is somewhat greater here, but you can react more quickly to adjustments as you only have to make them in one popup and not in different popups.

```
console.log("openDetailsDialog");
Popup_Details_None.setVisible(false);
Popup_Details_Title.applyText(Title);
if (G_DetailsFor === "None") {
Popup_Details_None.setVisible(true);
}
```

```
// Show Table
Popup_Details_Table_Employee.setVisible(false);
Popup_Details_Table_Entries.setVisible(false);
Table.setVisible(true);
// Current month
Popup_Details_Chart_CurrentPeriod.removeMeasure(G_Details_
CurrentMeasure, Feed.ValueAxis);
Popup_Details_Chart_CurrentPeriod.addMeasure(Measure, Feed.ValueAxis,
1);
// Previous month
Popup_Details_Chart_PreviousPeriod.
removeMeasure(G_Details_CurrentMeasure, Feed.ValueAxis);
Popup_Details_Chart_PreviousPeriod.addMeasure(Measure,
Feed.ValueAxis, 1);
// Average
Popup_Details_Chart_Average.removeMeasure(G_Details_CurrentMeasure,
Feed.ValueAxis);
Popup_Details_Chart_Average.addMeasure(Measure, Feed.ValueAxis, 1);
// Plan
Popup_Details_Chart_Planned.removeMeasure(G_Details_CurrentMeasure,
Feed.ValueAxis);
Popup_Details_Chart_Planned.addMeasure(Measure, Feed.ValueAxis, 1);
// Chart
Popup_Details_Chart.removeMeasure(G_Details_CurrentMeasure,
Feed.ValueAxis);
Popup_Details_Chart.addMeasure(Measure, Feed.ValueAxis, 1);
G_Details_CurrentMeasure
= Measure;
var selectedDims = ArrayUtils.create(Type.string);
var dims = Table.getDataSource().getDimensions();
if (dims.length > 0) {
for (var i=0;i<dims.length; i++){
CheckboxGroup_AllDimensions.addItem(dims[i].id,dims[i].description);
selectedDims.push(dims[i].id);
}
}
console.log(["selectedDims ", selectedDims]);
AllDimensions = selectedDims;
ScriptObject_Utils.setDimensionCheckboxes(Table);
Popup_Details.open();
```

Listing 3.15 Dynamic Call-Up of Popups

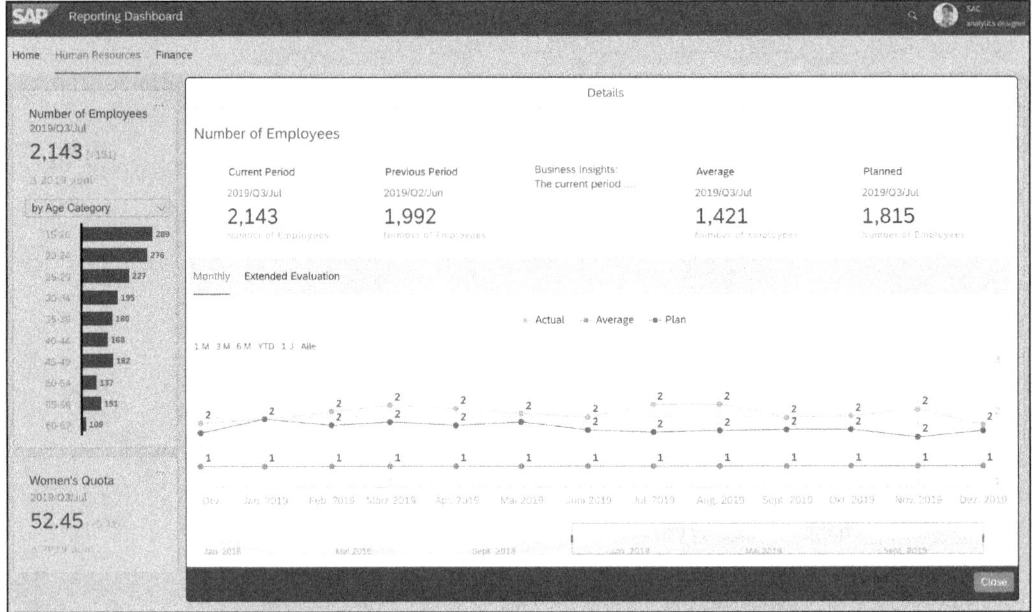

Figure 3.34 Details of Popup

In this section, we've presented two demo applications that demonstrate the possibilities of analytical applications. These and the other sample applications we've offered are very suitable for use in starting your own projects, or you can adopt parts of them for your own projects. However, the applications are also very suitable for discussion with specialist departments, as you can use them to demonstrate a finished application with very little up-front effort.

3.4 Integration of Predictive Analytics Functions

Smart functions

In addition to the classic BI functionalities, SAP Analytics Cloud offers so-called *predictive analytics functions*, which enable you to proactively identify problems and opportunities instead of just reacting to them. In essence, they help you predict the future. Various APIs are also provided in stories for these functions, which we'll take a closer look at now.

3.4.1 Smart Discovery

Let's start with the *smart discovery* function, which uses machine learning to analyze your data and help you put your data into a specific context, discover new data, or find previously undetected relationships among your data. We'll cover smart discoveryrelated APIs in the following sections.

SmartDiscovery

The SmartDiscovery API consists of two different methods:

- buildStory
- generateUrlToBuildStory

With the buildStory method, you have the option of creating a smart discovery URL that you can provide as a hyperlink in a dialog. An end user can then click that hyperlink to open the smart discovery story in a new browser window.

[«]

Availability on Mobile Devices

The smart discovery function is not available if you open it via a mobile device or a publication.

Another method of this API is the generateUrlToBuildStory method, which you can use to create a smart discovery URL. It doesn't open a dialog, but you can use the URL as a hyperlink on images or buttons, for example.

SmartDiscoveryDimesionSettings and SmartDiscoveryStructureSettings

The SmartDiscoveryDimesionSettings and SmartDiscoveryStructureSettings APIs contain various methods that you can use to create the dimensions and structure settings for a smart discovery. You can copy dimension filters of a dimension into your page filter and set various properties, and you can also show and hide the properties panel.

Listing 3.16 shows how you can use these methods. For example, you can place the following code on a button that is used to call the smart discovery function.

```
var ds = Chart_Forecast.getDataSource();
var members = ds.getMembers("Product_3e315003an");
var SDsetting = SmartDiscoveryDimensionSettings.create(ds, "Product_
3e315003an",[members[1]]);
SDsetting.
setIncludedDimensions(["Location_4nm2e04531", "Store_3z2g5g06m4"]);
SDsetting.setIncludedMeasures(["[Account_BestRunJ_sold].[parentId].&[
Gross_Margin]","[Account_BestRunJ_sold].[parentId].&[Discount]"]);
SmartDiscovery.buildStory(SDsetting);
```

Listing 3.16 Calling Up Smart Discovery

When you execute Listing 3.16, a window (shown in Figure 3.35) first opens in which you must confirm the start of smart discovery.

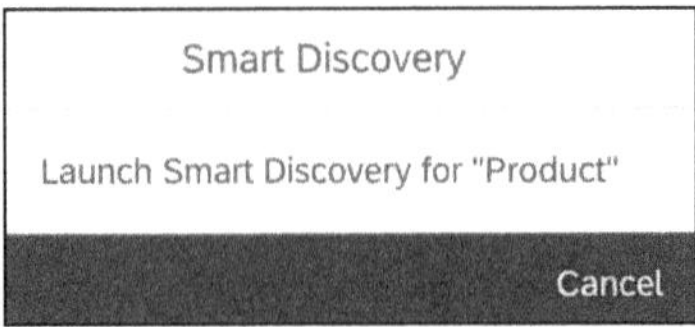

Figure 3.35 Dialog in Smart Discovery

Once you've started smart discovery, a new tab will open and you'll first be shown a loading animation (see Figure 3.36).

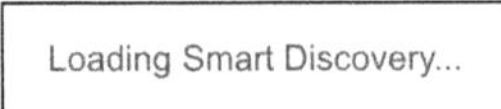

Figure 3.36 Loading Smart Discovery in New Window

Once the loading process has been completed, you can create your desired destinations and entities as well as advanced settings. This process will resemble the one depicted in Figure 3.37.

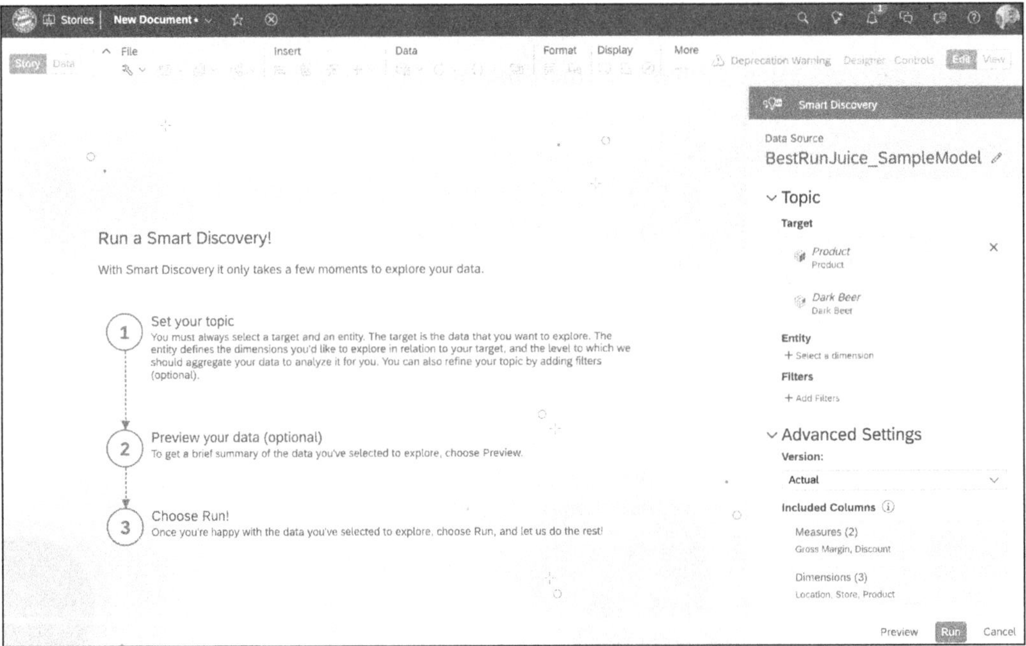

Figure 3.37 Executing Smart Discovery

Once you've made your settings, you can run your smart discovery function. You can see a possible result in Figure 3.38.

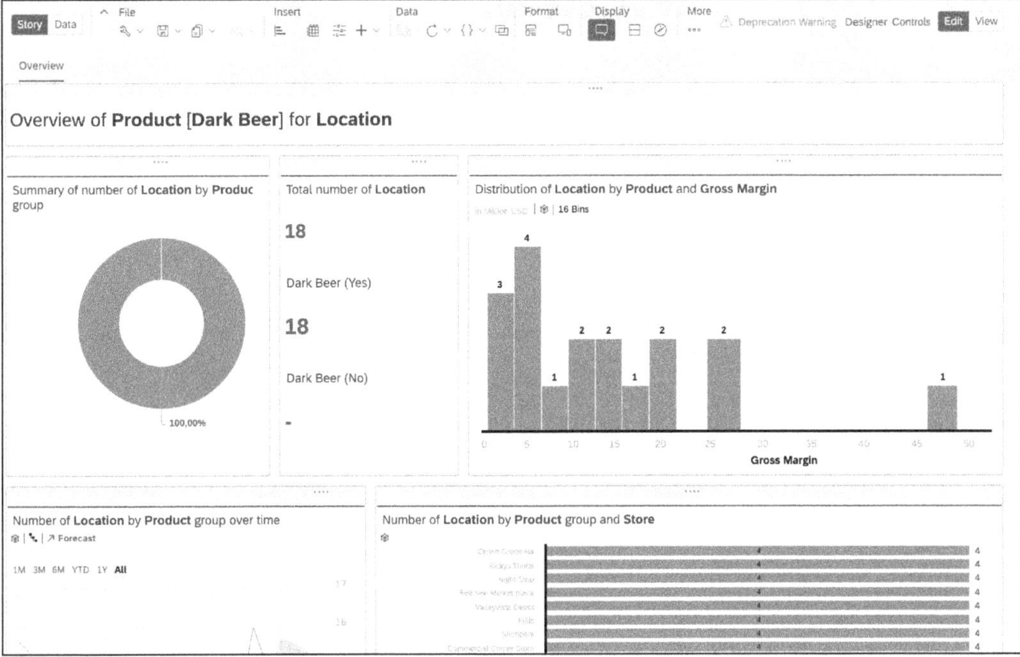

Figure 3.38 Result of Smart Discovery Function

3.4.2 Search to Insight

The *search to insight* function of SAP Analytics Cloud offers an interface in which you can query data using natural language. It lets you ask questions, get your answers from the data models, and start a new story.

Here, you can ask questions like these (depending on your data):

- How high was the turnover for Germany?
- Which item was sold most frequently in 2021?

[+]

Supported Languages

The search to insight function is currently only usable with English words.

The `SearchToInsight` API allows you to open and close the search to insight dialog. You can also insert your search to insight results into a chart or work with variables.

[+]

Availability on Mobile Devices

The search to insight dialog is currently not supported for mobile devices.

3.4.3 Data Change Insights

With the *data change insights* function, you can create snapshots of your data to compare them with one another. You can send these results by email or an app notification.

The API provides you with various methods for comparing the current status of the application with a snapshot or two snapshots. You can also save a snapshot or open the subscription dialog.

3.5 Best Practices

We recommend

There are some best practices for scripting that we can recommend to you. Basically, all best practices that apply to simple stories in SAP Analytics Cloud also apply here. We also recommend following the common code guidelines—this was not necessary or even possible in the classic stories, but you should observe the guidelines for the optimized stories. Remember to document your code sufficiently here too. If you pay attention to these points, you'll lay the foundation for a good and high-performance story.

Naming conventions

To help you maintain an overview of your story at all times, we highly recommend that you name your widgets from the outset. Think about a suitable naming convention for them (e.g., an area always starts with "B_") and remember to make names as descriptive and clear as possible. This will make it easier for you to select the right widgets in your code.

This all seems very simple and easy to keep track of at first, but as the complexity of a story increases, you'll quickly lose the overview. Of course, this recommendation applies not only to your widgets but also to other elements such as script variables.

Maintenance

In general, we highly recommend outsourcing your code to script objects and not storing the code in every event. This will make it easier for you to maintain your story, as you will be able to make changes in a central location and won't have to intervene in different events. This also increases the reusability of your code.

Basic framework

You should set up a kind of basic framework for your story, in which you incorporate widgets and functions that each of your dashboards must have. This has the advantage that you won't have to start from scratch with every new story and will therefore be able to develop your stories more quickly. You should also make sure that you have as few empty spaces as possible in your story. Always fill these with an area or flow layout area; this gives you more flexibility if you want to make changes to your story. For example, if you change the size of your header and there are five free charts under the header, you'll have to adjust the new spacing for each chart individually.

However, if you've grouped these five charts into one area, you'll only need to adjust the spacing of the area, and all five charts will then have the old spacing to the header again.

Own popups

Another tip we can give you is to create your own popups, which are also best placed in the basic structure of a story. Custom popups are particularly recommended if a story is not always accessed with the same screen size. The standard popup from SAP Analytics Cloud can only be set with fixed sizes, and we know from practical experience that the button for closing a popup is not visible on smaller-size screens and that the end user therefore won't have the option of closing the popup. We therefore recommend that you build your own popups using the area widget. Here, you can set the desired distances to the edge of the screen yourself and build your own header and footer area with additional areas and buttons. This way, you avoid this problem and have flexible popups.

For example, you can build popups that always have a 10% gap to the edge of the screen, and you can insert another area into this area that has a height of 5% and a 0% distance to the bottom edge of your other area. You can insert the same for your header, except that there should be no distance to the top edge. Then, place a button in your footer area and set the method that the area should be hidden to the `onSelect` event of this button. This is how to quickly and easily build your own popup (see Figure 3.39)!

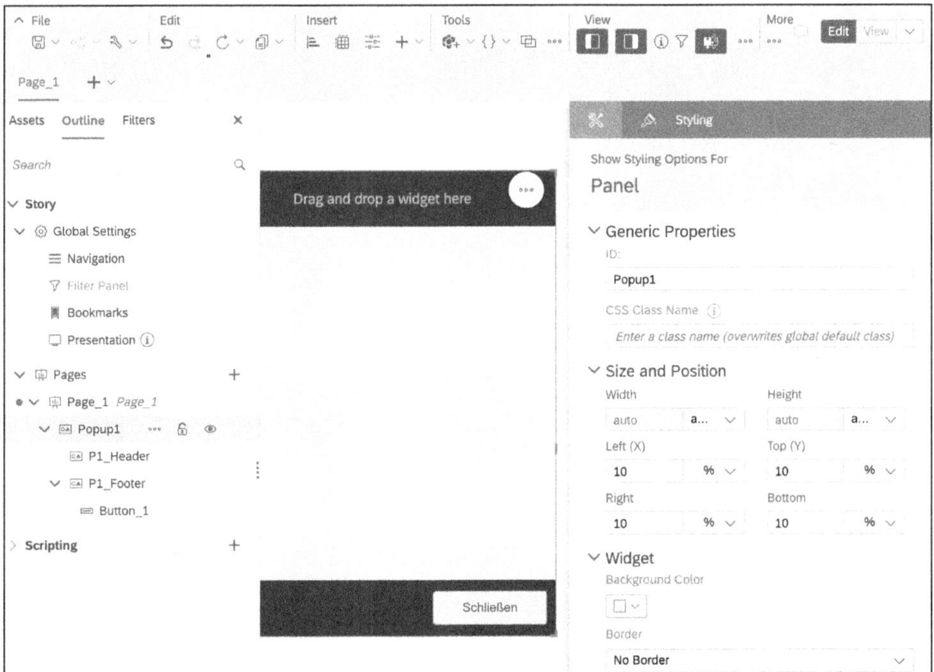

Figure 3.39 Create Your Own Popup

3.6 Summary

In this chapter, you gained a basic knowledge of the different widgets and APIs plus some detailed insights into them. After that, you learned how to develop a story and integrate predictive functions into it. Finally, you got detailed information from some stories with scripting logic to help you learn how to use stories.

In the next chapter, you'll learn to extend your own BI stories with scripting.

Chapter 4
Development of Planning Applications

In this chapter, you'll learn how to create a planning application using the various widgets and APIs.

What to expect

In the previous chapter, we showed you how to use SAP Analytics Cloud to create stories with scripting for reporting purposes. In this chapter, we show you which features you need to consider for planning stories. In Section 4.1, "Widgets," and Section 4.2, "Application Programming Interfaces," we present the most important widgets and APIs available for planning stories, using simple code examples (just as we did in the early sections of Chapter 3). We also look at the additional possibilities of planning stories compared to BI stories. In Section 4.3, "Data Connection for Writing Back Data," we explain why writing data back to the source systems via an OData interface plays an important role in planning stories. Then, in Section 4.4, "Developing a Story with Scripting for Planning," we develop an exemplary planning story step by step. Finally, in Section 4.5, "Best Practices," we present best practices for the development of stories with scripting for planning purposes, using a number of examples. The examples we show are based on the business content package *SAP Operational Workforce Planning*. This business content is available to you free of charge, and it contains planning models, several analytical applications, and predefined planning objects (which are data actions and multi actions).

[+]

Business Content

SAP provides *business content* free of charge. This consists of ready-made content packages for various industries and business areas. In addition, SAP makes data models and all other necessary objects available to you so that you can use the content directly. You can activate business content via the *content network*.

4.1 Widgets

Widgets for planning scenarios

In the following section, we'll introduce you to widgets that are particularly relevant in the planning context. You can use these widgets to make your stories interactive for users.

You can find the widgets that we'll discuss via **Assets** in the left-side panel under **Planning Actions** and **Others** (see Figure 4.1).

Figure 4.1 Widgets Relevant for Planning Functionality

Widget in focus

We'll introduce the following widgets in the next sections:

- Comment
- Value driver tree
- Data action trigger
- Multi action trigger
- Business planning and consolidation planning sequence trigger

4.1.1 Comment

In addition to the known possibilities of inserting comments directly into tables or chart widgets, you can use the *comment* widget (see Figure 4.2).

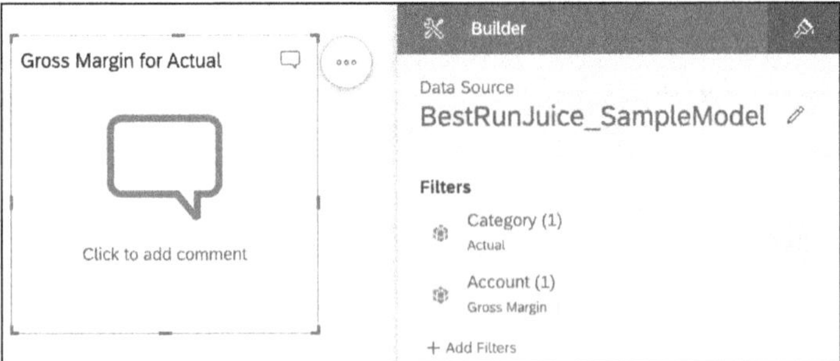

Figure 4.2 Comment Widget without Comments

You can use the comment widget to add data model or data point comments. Depending on the filter setting, both are possible. As you can see in Figure 4.3, comments within the widget are displayed in list form and sorted by recency; the most recent comment appears first.

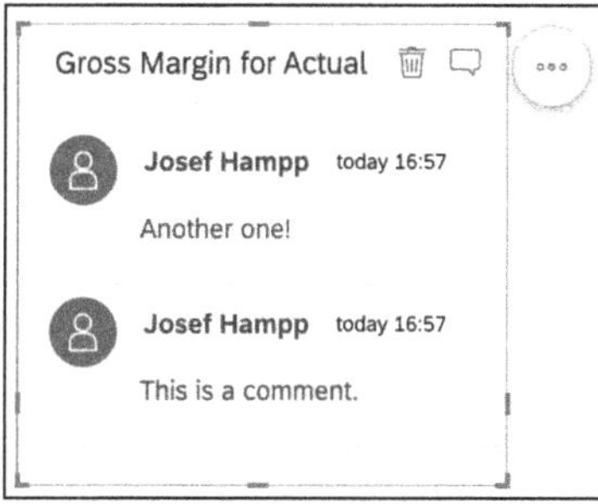

Figure 4.3 Comment Widget with Comments

Storage of Comments

Comments within the comment widget are saved at model level, and they are independent of the story.

Commenting with the comment widget

You can add new comments using the speech bubble icon at the top right, which opens an input area where you can type comments. Figure 4.4 shows you the available formatting options. In addition to the standard font, you can emphasize special features by writing in bold or italics and by underlining or crossing out words. You can add numbered and unnumbered lists, use hyperlinks, and change the font color, and you can also utilize the basics of a proper text editor.

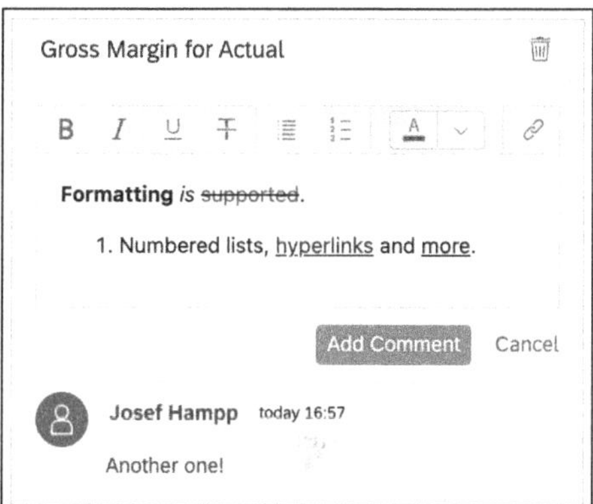

Figure 4.4 Formatting Options for Comment Widget

[!]

Unsupported Functionalities

The use of linked analyses is not supported by the comment widget.

Comments are also not supported for live data models or blended models. In such cases, you can't add comments via the comment widget. Exceptions are the SAP Business Warehouse (SAP BW) live and SAP Business Planning and Consolidation (SAP BPC) live data models.

4.1.2 Value Driver Tree

Driver-based planning approach

You can use the *value driver tree* widget to display so-called value driver trees in SAP Analytics Cloud. With a value driver tree, you can pursue driver-based planning approaches, which seek to identify the most important value drivers and model their impacts on the most important key performance indicators (KPIs). With this approach, you can look at the entire value chain of your company, instead of isolated key figures. Figure 4.5 shows an example of a value driver tree.

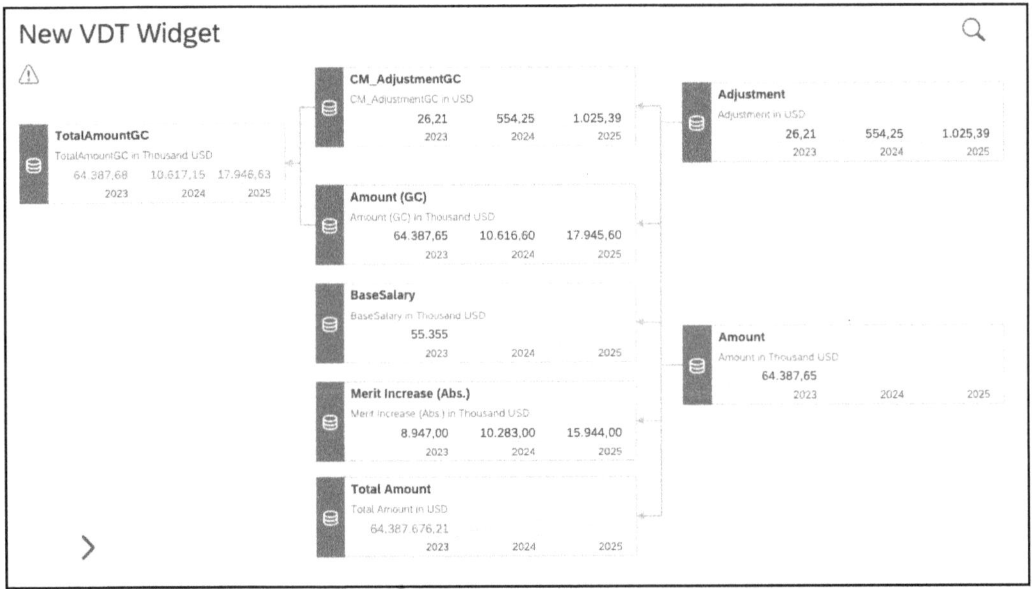

Figure 4.5 Value Driver Tree

The individual drivers of a value driver tree are divided into two categories:

- Internal factors (e.g., personnel costs, production costs)
- External factors (e.g., customs duties, raw material prices)

The aim is to visualize the relationships between individual factors and important key figures such as total sales and net profit in a model.

Comparing value driver trees with tables

Like tables, value driver trees show numerical values and allow you to customize your plan values (e.g., with the help of assisted data entry; see Figure 4.6). Unlike tables, value driver trees allow you to display links between the key figures and hierarchies in the data model.

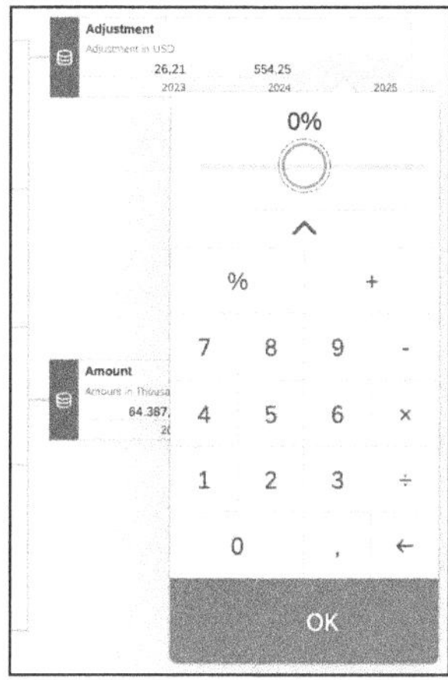

Figure 4.6 Value Driver Tree Data Entry

Figure 4.5 shows a value driver tree for human resources data. In this example, one key figure is shown in each node, and the visible period is defined from 2023 to 2025. As the creator of a value driver tree, you can define these settings yourself. You can define which key figures or calculations are displayed in a node, determine filter settings, and expand or collapse node groups. A diagram is created based on your settings.

Advantages of a value driver tree

With these visualization options, the value driver tree gives you an optimal view of your most important key figures. You can compare different time periods with each other and thus easily enter into medium- to long-term planning.

Prerequisites for Generating Value Driver Trees

The creation of a value driver tree widget is license independent. Both planning and BI users can create value driver trees, but only planning users can make changes to a version and publish them.

4.1.3 Data Action Trigger

Automating planning steps

Data actions are structured actions that allow you to automate repetitive and time-consuming steps in a planning cycle. You can use data actions to copy or paste data, perform scripted calculations or allocations, and publish versions of data.

Before you can integrate a data action into your story, you need to create it, which you do in a special editor. You can open the special editor in the navigation bar by clicking the **Data Actions** menu item.

Figure 4.7 shows the data action SAP__HR_BPL_IM_CALCULATE_DELTA_VALUES. This data action consists of only one step: an *extended formula step*, which you can use to map more complex calculations or transformations. You can define these in script language or model them visually (see Figure 4.8).

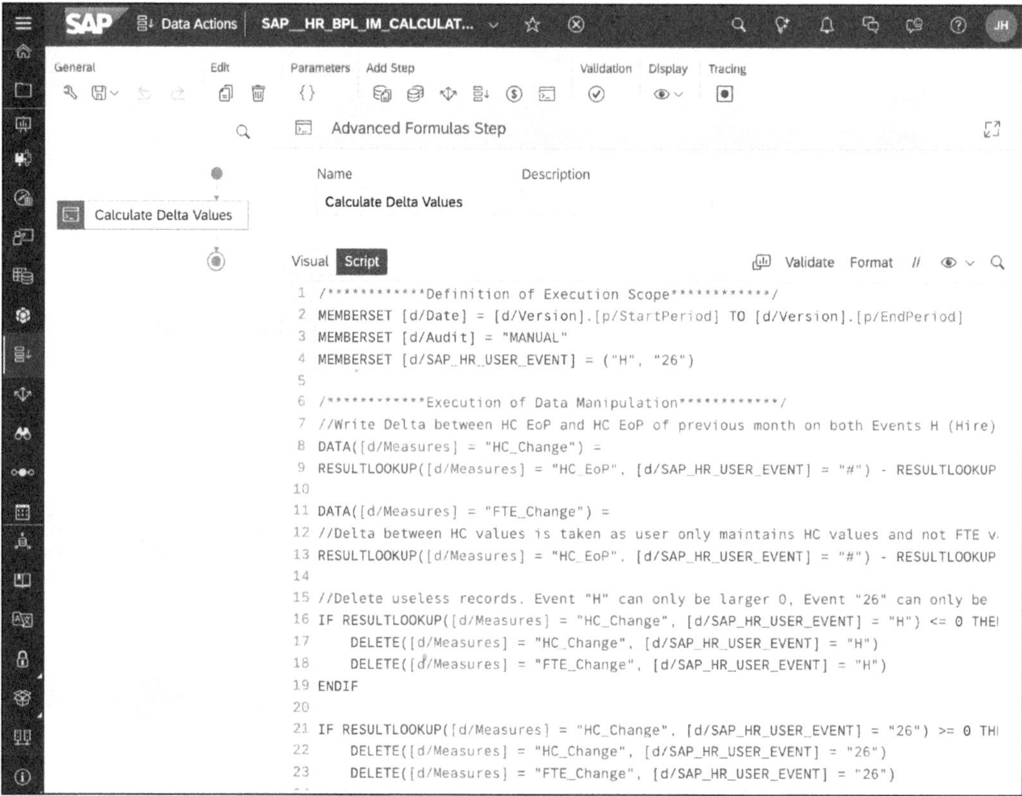

Figure 4.7 Example of Data Action

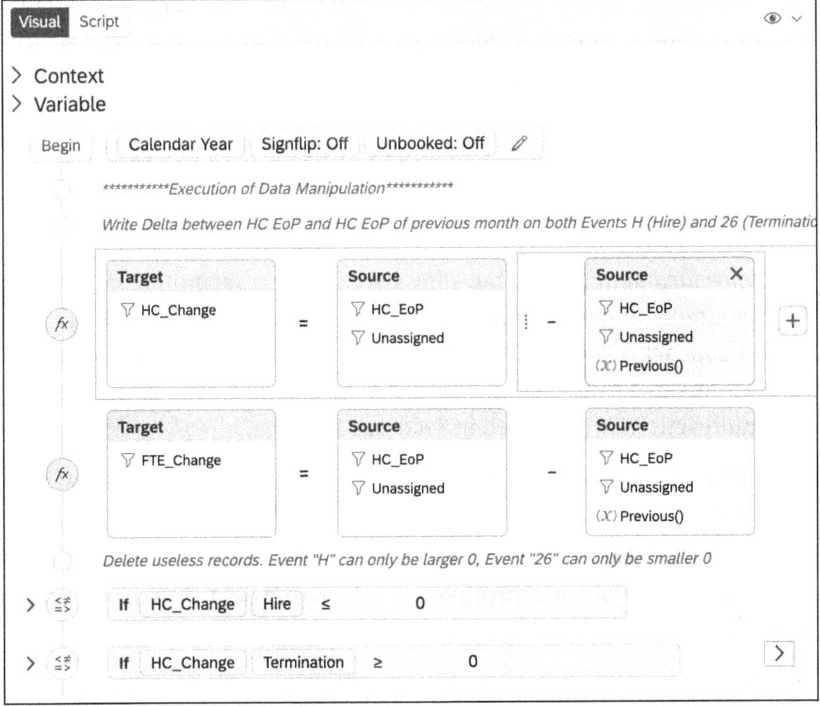

Figure 4.8 Visual Representation of Data Action

In addition to the extended formula step, there are other standard steps that you can use to create a data action. You can add these steps via the header menu of the data action. Figure 4.9 shows the various icons for adding new steps.

Figure 4.9 Menu for Creating New Data Action Steps

From left to right, you can use the following steps:

- Copy step
- Cross-model copy step
- Allocation step
- Embedded data action step
- Currency conversion step
- Advanced formulas step

Data action steps

The *copy step* allows you to copy data using filter and aggregation settings. In this way, you can copy data from the previous year to the current year.

The *cross-model copy step* offers similar functions, and its main difference from the copy step is that you must take the two different models (the source planning model and the target planning model) into account in the cross-model copy step.

With the *allocation step*, you can use allocations as part of your data action. You can allocate values of a source dimension to a target dimension via direct allocation or with the help of value factors.

The *embedded data action step* allows you to execute another data action from your existing data action. You therefore have the option of creating nested data actions with this step, and you can also model a sequence of data actions in a multi action. The exact distinction between data actions and multi actions is explained in Section 4.1.4.

The *currency conversion step* allows you to convert key figures into another currency. You can define conversion rules and the write mode.

You've already seen the *extended formula step* in Figure 4.7, and you use it to create a script for transforming and calculating data. This allows you to perform consolidation and planning tasks.

Now that you've defined a data action, you need to make it executable in your story. You can do this by using the *data action trigger* widget, with which you can create a clickable object for end users. Clicking the object executes the stored data action, and Figure 4.10 shows what this can look like.

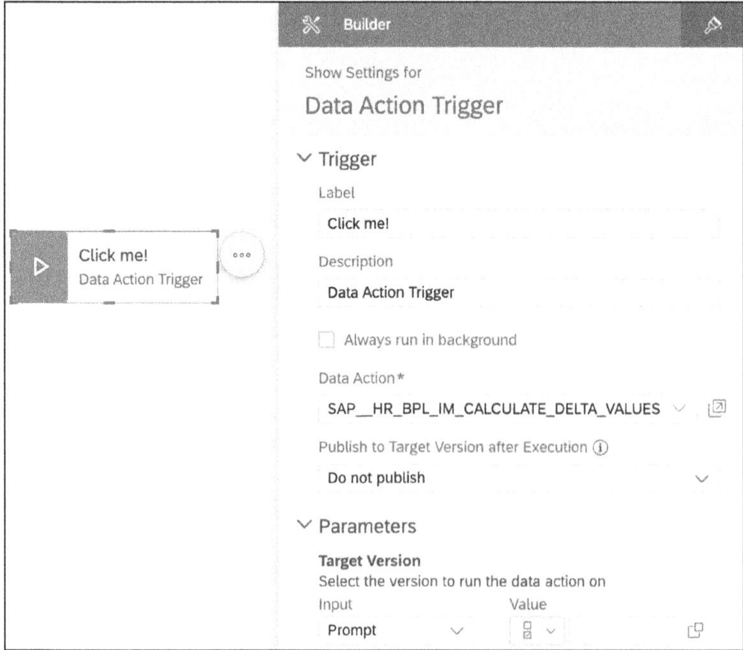

Figure 4.10 Data Action Trigger

4.1.4 Multi Action Trigger

Automating planning processes

You use *multi actions* to control planning and forecasting processes across models and versions. With multi actions, you can define the structured execution of data actions, forecasting, and version management steps. This allows several data actions to be executed in succession, forecast scenarios to be realized, or versions to be published. Of course, you can also use combinations of these steps.

Figure 4.11 shows the `SAP__HR_BPL_IM_WFP_SmartPredict_PL1` multi action, which is also part of the SAP Operational Workforce Planning business content. As you can see, the multi action simply defines a sequence of different steps to be executed. A distinction is made between six different step types, and you can add these via the menu items in the header.

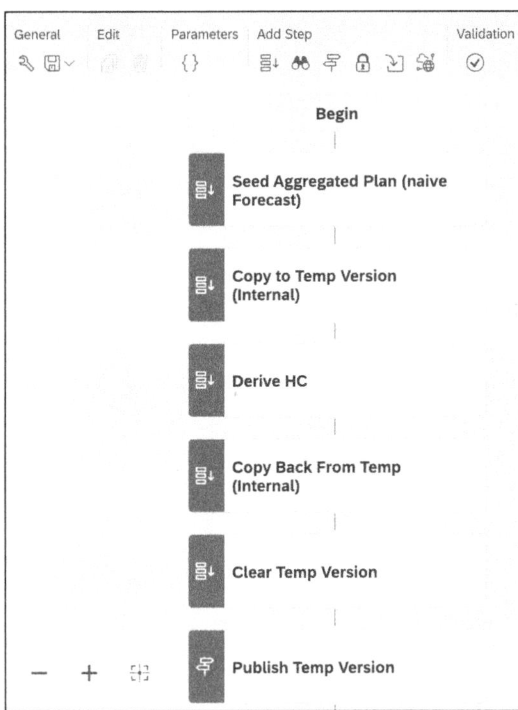

Figure 4.11 Example of Multi Action

Figure 4.12 shows the **Add Step** section for creating new *multi action steps*. You can create the following objects using the icons in this section:

- Data action step
- Predictive step
- Version management step
- Data locking step

- Data import step
- API step

Figure 4.12 Section for Creating New Multi Action Steps

Multi action steps

You can add a *data action step* by clicking the first icon on the left. This allows you to execute individual data actions within your multi action. You can also parameterize them, which means that you can use fixed or dynamic parameter values to influence the execution of the data actions.

You can also use a multi action to execute time series forecast scenarios. To do this, you can use the *predictive step* by clicking the binoculars icon. In this way, a forecast model is trained using your data and is then written back to a planning version.

The *version management step*, which you can add by clicking the signpost icon, allows you to publish a version from the planning model.

You can use *data locking steps*, which you can add by clicking the lock icon, to set data locks directly via the multi action trigger. Setting data locks protects your data from unwanted changes.

The *data import step*, which you can add by clicking the second icon from the right, allows you to use data from other sources in your multi actions. By using multiple data import steps, you can use data from several different sources.

You can add an *API step* by clicking the globe icon on the right. With an API step, you can integrate on-premise and cloud-based applications via an `http` API.

[!]

Forecast Capabilities of Multi Actions

You can't use all algorithms of SAP Analytics Cloud's predictive scenarios with the predictive step. In a predictive step, you can only call time series forecast scenarios.

After defining a multi action, you can make it executable via a widget in your story. Figure 4.13 shows you how to use the *multi action trigger* widget. It creates a clickable object with which users can start the multi action.

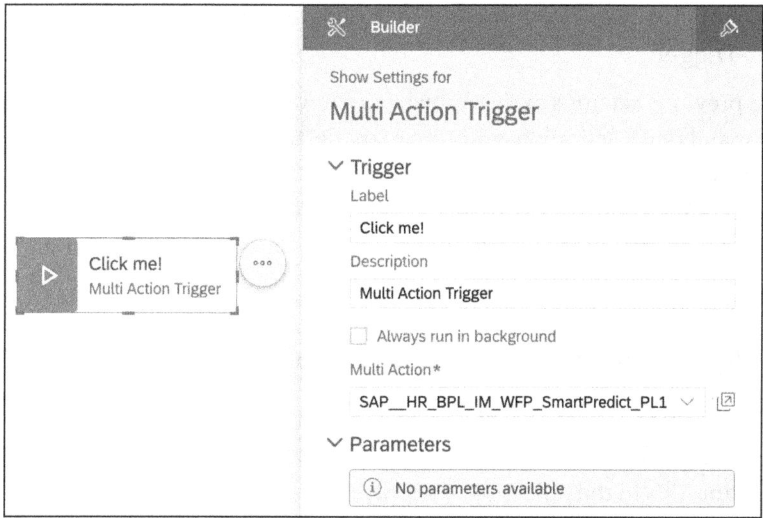

Figure 4.13 Multi Action Trigger

4.1.5 Recommendation for the Use of Data Actions and Multi Actions

Now that we've introduced data actions and multi actions, you'll notice that these two actions are very similar in some aspects. In particular, the functionality of the embedded data action steps allows you to copy the behavior of multi actions. With both objects, you can execute several data actions sequentially. This raises the logical question of when you should use data actions and when you should use multi actions when creating stories.

Data actions or multi actions

The answer is that you should use data actions if you want to perform a large number of transformations and calculations on planning data, and you can also use data actions to edit only one version of your planning model at a time. Multi actions, on the other hand, are designed to simplify processes for users, and the possible use cases for multi actions go far beyond those of data actions. Therefore, you can use multi actions to work with different target models and target versions. For example, you could plan with several versions at the same time and fill them with data. You can also use multi actions to work with time series forecasts and transfer them to your planning model, or you can use multi actions to publish data within a process and then carry out further steps. As you can see, multi actions cover a wider range of applications than data actions in the context of planning processes.

4.1.6 SAP Business Planning and Consolidation Planning Sequence Trigger

In the previous sections, we've shown you how to define and execute data actions and multi actions. In addition to modelling planning processes, SAP Analytics Cloud lets you integrate with *SAP BPC*. With the *business planning and consolidation planning sequence trigger* widget, you can trigger planning sequences that you've created in SAP BPC via SAP Analytics Cloud.

4.2 Application Programming Interfaces

SAP Analytics Cloud also provides many APIs in the planning context that you can use when creating a story with scripting. We present the most important APIs in the following sections.

4.2.1 DataAction

Data actions

In Section 4.1.3, we explained how to use a widget to execute data actions. This option is also available with the `DataAction` API, with which you can use the following methods:

- `Execute`
  This method executes the data action. Note that this is a blocking method, meaning that it prevents further execution of the application script until the data action is completed.
- `executeInBackground`
  This method executes the data action as a nonblocking operation.
- `getExecutionProgress`
  This method returns the current status of the data action.
- `getParameterValue`
  This method returns the value of the parameter.
- `isAllMembersSelected`
  This method specifies whether the all members parameter has a value.
- `setAllMembersSelected`
  This method sets all members as the parameter value.
- `setParameterValue`
  This method sets the value of the parameter.

[+]

Optimal Use of Data Actions

Since calling the `execute` method blocks the story, you should only define data actions that are executed quickly. If you use data actions that block the application for a longer period of time, it will have a negative impact on the user experience.

In the following sections, we'll show you how to create a new data action scripting element and its configuration. After that, we'll describe the most important methods of the DataAction API.

Creation of a Data Action Scripting Element

To use the DataAction API, you must first define the data action as a scripting element in your story. As with the data action trigger widget, you must first create the data action separately.

You can create a new data action scripting element in your story via the **Outline** area in the left side panel. To do this, open **Outline** and click the **+** icon next to **Scripting**. Here, you'll find **Data Actions** as an item, and you can then create a new data action by selecting **Data Actions** in the popup window (see Figure 4.14).

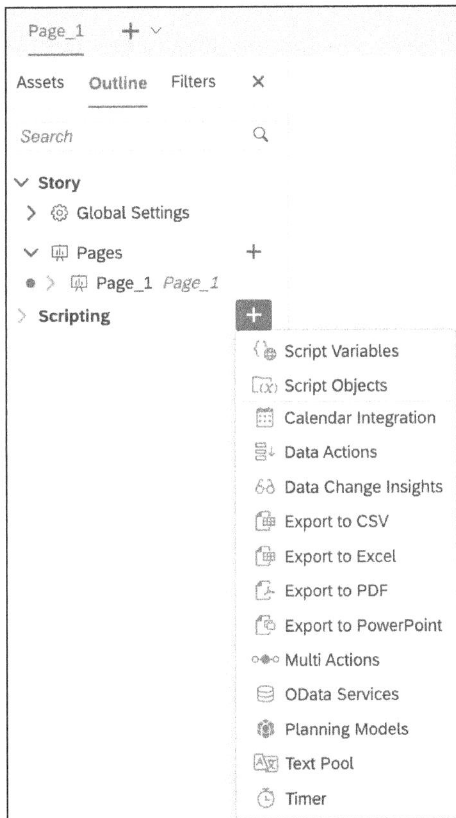

Figure 4.14 Adding Data Action Scripting Element

After you create a new data action scripting element, a **Data Actions** item is added to the **Scripting** section and the newly created scripting element is displayed. In Figure 4.15, it's labeled **DataAction_1**.

Figure 4.15 Newly Created Data Action Component

However, the data action you've created is still an empty object. It doesn't yet have a reference to an existing data action, so to add the reference, you must define the corresponding data action. To do this, click on the newly created scripting element, and a panel with the heading **Data Action Configuration** will open on the right-hand side.

The panel consists of two areas. In the upper area, you can select the data action to be executed, define a follow-up action, and decide whether the target version should be published automatically. In the lower area, you can define the parameters required for the data action.

For this example, select data action **SAP__HR_BPL_IM_CALCULATE_DELTA_VALUES**. This is part of the SAP Operational Workforce Planning business content package (see Figure 4.16).

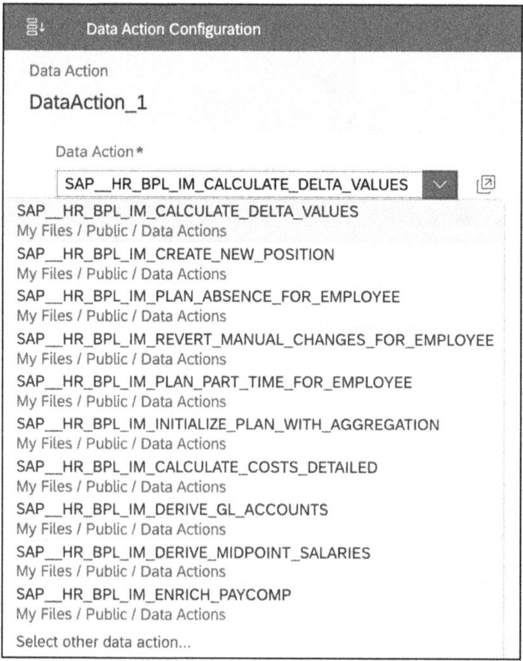

Figure 4.16 Selection of Data Action to Be Executed

Once you select a data action, the **Parameters** area will be updated (see Figure 4.17). For the data action selected in our example, you must still define the target version as a fixed value.

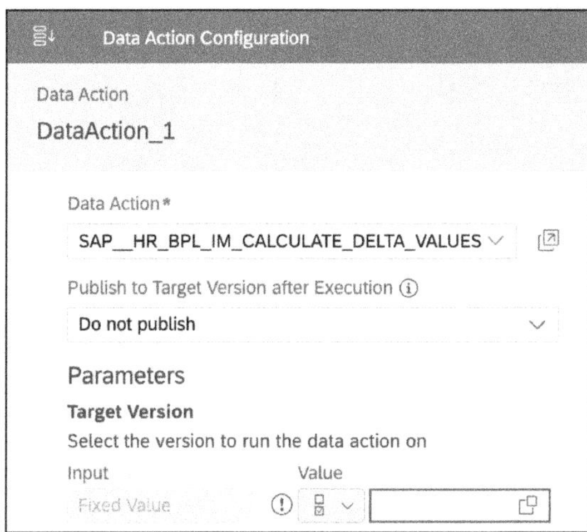

Figure 4.17 Definition of Target Version

You can select the target version via an additional popup window (see Figure 4.18).

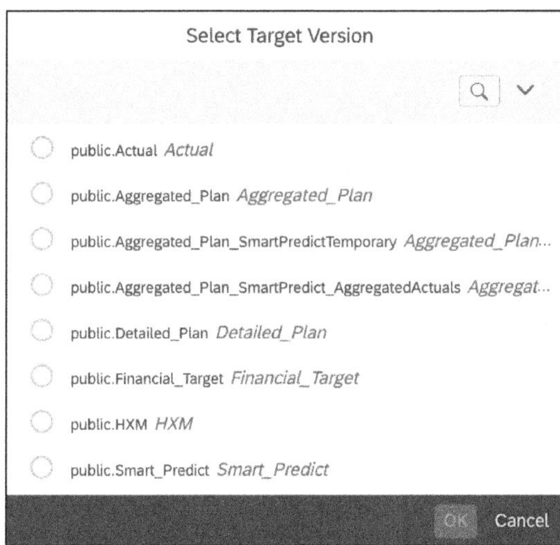

Figure 4.18 Input Option for Target Version

Once you have determined all the settings, you can use the data action as a script element.

DataAction Application Programming Interface Examples

To be able to execute our examples, you need to create a new button and add an **onClick** event to it. You can create the event via the context menu of the button widget as shown in Figure 4.19.

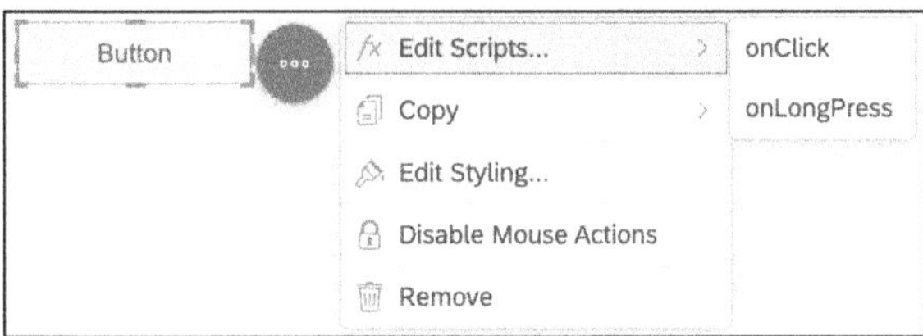

Figure 4.19 Creating Event for Button

In the previous section, we defined a parameter in the data action configuration, so we read this in the first step using the `getParameterValue` method. You can use this script to create the new `parameterValue` variable, which contains the `DataAction_1` parameter with the `TargetVersion` ID. You can use the `console.log` command to output this variable, as you can see in Listing 4.1.

```
var parameterValue = DataAction_1.getParameterValue(‘TargetVersion’);

console.log(parameterValue);
```

Listing 4.1 Log of Parameter Value

As you can see in Figure 4.20, the result of the `getParameterValue` method is a JSON object or an object of the `DataActionParameterValue` type.

```
▾ {type: 'Member', members: Array(1)} i
  ▸ members: ['public.Detailed_Plan']
    type: "Member"
  ▸ [[Prototype]]: Object
```

Figure 4.20 Log of Parameter Value

You can use the loop in Listing 4.2 to resolve this object and reuse the individual parameter value members.

```
var parameterValue = DataAction_1.getParameterValue('TargetVersion');

console.log(parameterValue);

var memberParameter = cast(Type.DataActionMemberParameterValue,
parameterValue);
var memberIds = memberParameter.members;
```

```
for (var i = 0; i < memberIds.length; i++) {
        console.log("parameter value: "+memberIds[i]);
}
```

Listing 4.2 Log of Parameter Value Members

In this case, the parameter values are again output via the console. The target version is `public.Detailed_Plan` (see Figure 4.21).

```
▾ {type: 'Member', members: Array(1)} i
  ▸ members: ['public.Detailed_Plan']
    type: "Member"
  ▸ [[Prototype]]: Object
parameter value: public.Detailed_Plan
```

Figure 4.21 Log of Parameter Value Members

In addition to reading a parameter value, you can set it by entering the ID of the parameter and its value. You can also use the following script to overwrite the target version and set a different version:

```
DataAction_1.setParameterValue("TargetVersion","public.New_Plan");
```

To check the execution, you can run the `getParameterValue` method again and display the result in the console. Since you now know that there's only one object, you can replace the loop and directly retrieve the first object in the array (see Listing 4.3).

```
var parameterValue = DataAction_1.getParameterValue('TargetVersion');

console.log(parameterValue);

var membersParameter = cast(Type.DataActionMemberParameterValue,
parameterValue);
var memberIds = membersParameter.members;
for (var i = 0; i < memberIds.length; i++) {
        console.log("parameter value: "+memberIds[i]);
}

DataAction_1.setParameterValue("TargetVersion","public.New_Plan");

parameterValue = DataAction_1.getParameterValue("TargetVersion");
membersParameter = cast(Type.DataActionMemberParameterValue,
parameterValue);

console.log("new target version: "+membersParameter.members[0]);
```

Listing 4.3 Set New Target Version

Figure 4.22 shows the execution of the code from Listing 4.3 and the resulting console output.

```
▼ {type: 'Member', members: Array(1)} i
  ▶ members: ['public.Detailed_Plan']
    type: "Member"
  ▶ [[Prototype]]: Object
parameter value: public.Detailed_Plan
new target version: public.New_Plan
```

Figure 4.22 Log of New Target Version

These examples have shown you how to set and read the parameters of a data action. Next, we look at how to execute a data action.

Executing a data action is simple because of the execute method, as shown in Listing 4.4. The method doesn't require any further parameters, but we recommend that you collect the return value of the method, as this tells you the status in which the data action was completed.

```
var parameterValue = DataAction_1.getParameterValue('TargetVersion');

console.log(parameterValue);

var membersParameter = cast(Type.DataActionMemberParameterValue,
parameterValue);
var memberIds = membersParameter.members;
for (var i = 0; i < memberIds.length; i++) {
        console.log("parameter value: "+memberIds[i]);
}

DataAction_1.setParameterValue("TargetVersion","public.New_Plan");

parameterValue = DataAction_1.getParameterValue("TargetVersion");
membersParameter = cast(Type.DataActionMemberParameterValue,
parameterValue);

console.log("new target version: "+membersParameter.members[0]);

var response = DataAction_1.execute();
console.log(response);
```

Listing 4.4 Executing Data Action

The status of a data action is returned as the DataActionExecutionResponse type. Figure 4.23 shows you the object in the console.

```
▼ {type: 'Member', members: Array(1)} i
  ▶ members: ['public.Detailed_Plan']
    type: "Member"
  ▶ [[Prototype]]: Object
parameter value: public.Detailed_Plan
new target version: public.New_Plan
▼ {status: 'Error'} i
    status: "Error"
  ▶ [[Prototype]]: Object
```

Figure 4.23 Log of Executed Data Action

In this example, the data action has been completed with the `Error` status, which means that the data action has been ended with errors. This can be explained by the fact that the new `New_Plan` target version doesn't exist and the data action can only be executed successfully if the target version is a valid version of the model. There are also other statuses:

- `Accepted`
- `Canceled`
- `Queued`
- `Running`
- `Success`

A data action has the `Accepted` status if the execution has been accepted.

A data action has the `Canceled` status if it has had to be cancelled. This can have various causes.

If a data action has been accepted but can't yet be executed because another data action is being processed first (i.e., if it's in a queue), then the data action has a `Queued` status.

If a data action has been executed, it has the `Running` status.

If a data action has been successfully completed, it has the `Success` status.

4.2.2 ODataService

OData

The `ODataService` API provides you with various methods for executing and reading OData services:

- **`executeAction`**
  You can use this method on existing OData services scripting objects. As an argument to the method, you can specify an OData action that is then called.

- `getEntitiesFromEntitySet`
  You can use this method to read the data of an OData endpoint. By specifying a name, you can restrict the call to an entity set. You can also use these other optional filter terms to specify your request:
  - `filter`
  - `orderby`
  - `select`
  - `skip`
  - `top`

4.2.3 MultiAction

Multi Actions

In Section 4.1.4, we explained how to use a widget to execute multi actions. You can also use the `MultiAction` API to execute multi actions via scripting, and you can use the following methods with the API:

- `execute`
  This method executes the multi action. Note that this is a blocking method, meaning it blocks further execution of the script until the multi action is completed.
- `executeInBackground`
  This method executes the data action as a nonblocking operation.
- `getParameterValue`
  This method returns the value of the parameter.
- `setParameterValue`
  This method sets the value of the parameter.

4.2.4 Planning

Planning APIs

The APIs of the `Planning` type library use its methods to extend various widgets that are used for planning, as described in the following sections.

BpcPlanningSequence

BPC planning sequence

The `BpcPlanningSequence` API extends the *business planning and consolidation (BPC) planning sequence trigger* widget, and you can use the `onBeforeExecute` event to read out whether a planning sequence is being executed. If you click on the BPC planning sequence trigger widget and this event returns `true` or no value, the planning sequence is started. If the result is `false`, the planning sequence is ignored.

This API also contains three additional methods:

- `Execute`
  This method executes the business planning and consolidation planning sequence.
- `getBpcPlanningSequenceDataSource`
  This method returns the data source of the business planning and consolidation planning sequence.
- `openPromptDialogue`
  This method opens the prompt dialog window of the business planning and consolidation planning sequence.

DataActionTrigger

Data Action Trigger

The `DataActionTrigger` API extends the basic widget APIs of the data action trigger widget and adds the `onBeforeExecute` event. This event is executed when users of the story click on a data action trigger widget. You can use this method to see whether a data action should be executed. If this method returns `true` or no value, the data action is triggered, and if it returns `false`, the data action is not executed.

Data Locking

The following methods of the `DataLocking` API provide you with functions for handling data locks:

- `getState`
  You can use this method to query the data lock status of a data cell.
- `setState`
  You can use this method to set the status of a data cell. If the execution is successful, `true` is returned as the result, and if it is unsuccessful, `false` is returned.

Data locks allow you to protect certain data from changes. This is a great advantage in more complex planning scenarios if you only want to make certain data areas available to your users for modification.

Activating data locks for a planning model

Firstly, we would like to show you how to activate data locking for a planning model. In this example, we are using the **SAP__HR_BPL_IM_WORKFORCE** planning model from the SAP Operational Workforce Planning business content package. To activate or deactivate data locking for a model, you must open the model preferences, which you can do if you are in the modeler. In the **Model Structure** or **Calculations** workspace, open the model preferences via the tool (wrench) icon in the **General** menu area (see Figure 4.24).

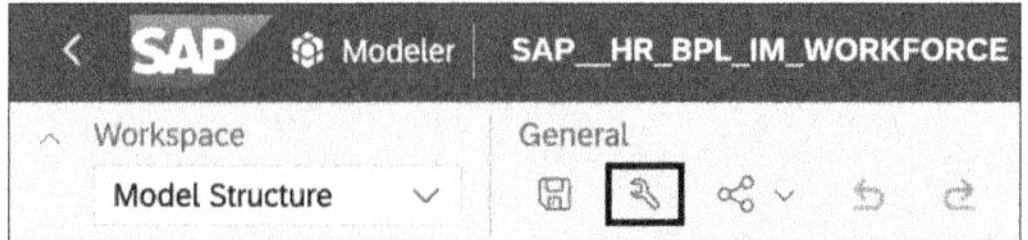

Figure 4.24 Opening Model Preferences

The model preferences are displayed in a popup window that is divided into several areas. To configure the data locks, open **Access and Privacy** (see Figure 4.25). In the **Security** section, you'll find **Data Locking** as an item, and you can use the switch there to set the data lock for a model. You'll then be able to lock individual data slices of the model.

In general, there are two ways to create data locks:

- Manual data locking
- Scripted data locking

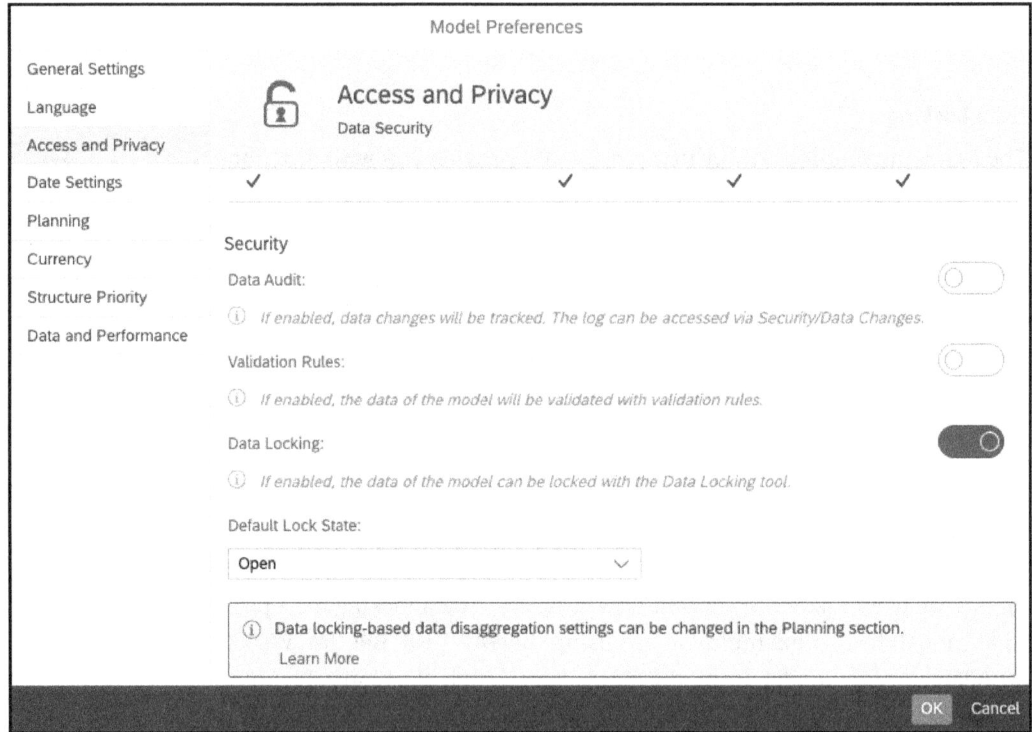

Figure 4.25 Enabling Data Locking via Model Preference

Set data locks manually

If you want to create a data lock manually, you can do it directly in the modeler or within a story by using table widgets. You need the appropriate privileges for both options.

First, let's look at managing the data lock via the modeler. To do this, open the planning model in the modeler. You can't adjust the data lock if you are in the open data management view, so open the model in the **Model Structure** or **Calculations** workspace. Figure 4.26 shows you the header menu of the modeler, and in the **Data** section, you'll find the lock icon, which you can use to open the data lock configuration.

Figure 4.26 Icon for Opening Data Lock Configuration

Figure 4.27 shows the menu for configuring the data lock, which opens in a new window. The process consists of two steps. In the first step, you select the **Driving Dimensions** and define the **Data Locking Owners**. *Driving dimensions* are the dimensions that are required to define the data lock to be locked. You can define the data lock ownership for each dimension, and then, you can assign the rights to individual users.

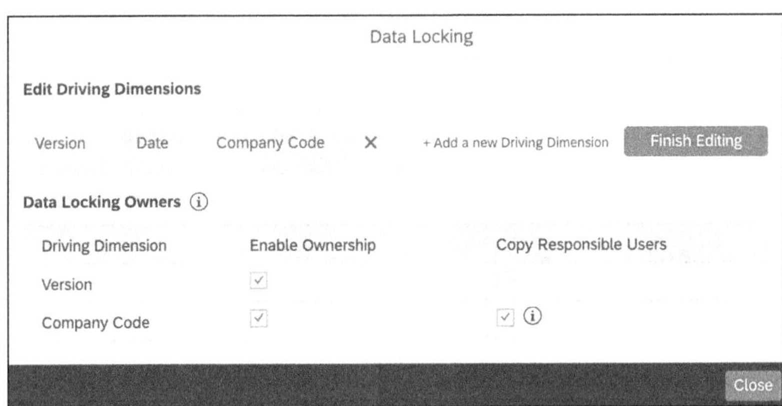

Figure 4.27 Adding Driving Dimensions and Defining Data Locking Owners

Once you've set this option, you can define the data slice in the second step by setting filters for the individual driving dimensions. Based on these filters, the data slice is then displayed in a table. Figure 4.28 shows you how you can define the lock status for individual data cells. You can choose between the following statuses:

- **Open**
- **Restricted**
- **Locked**

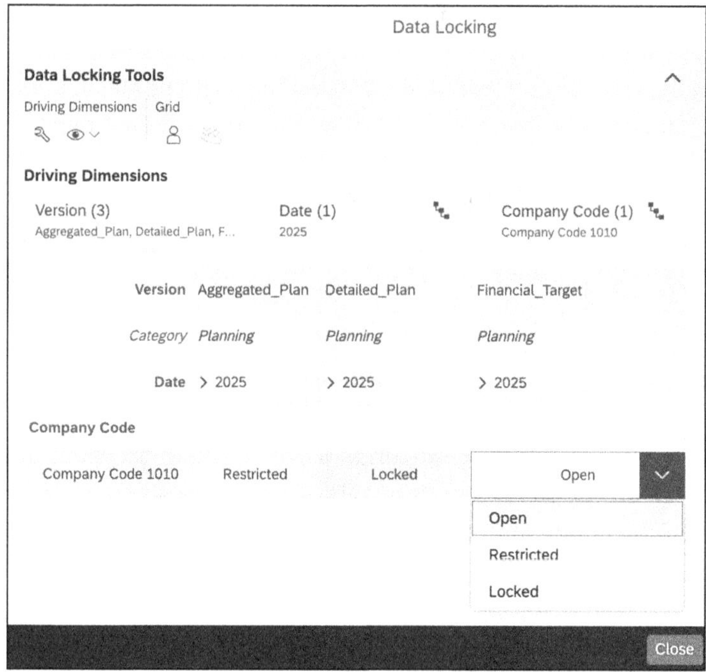

Figure 4.28 Setting Data Lock on Data Slice

Locking states

Open indicates that the data values for this data slice are unlocked, and all users who have the corresponding privileges to enter values for the planning model can edit these data values.

Restricted indicates that only users with lock ownership can edit the data values of the data slice.

Locked indicates that no one can edit the data values.

Hierarchy Behavior

Subordinate (child/lower level) elements of a dimension take on the lock status of the parent element.

To assign ownership to users of a restricted data slice, you must select it and then open the menu for defining the owners via the header menu in the **Grid** section (see Figure 4.29).

Figure 4.29 Icon to Open Ownership Assignment

Figure 4.30 shows this menu. You can add owners for each driving dimension here.

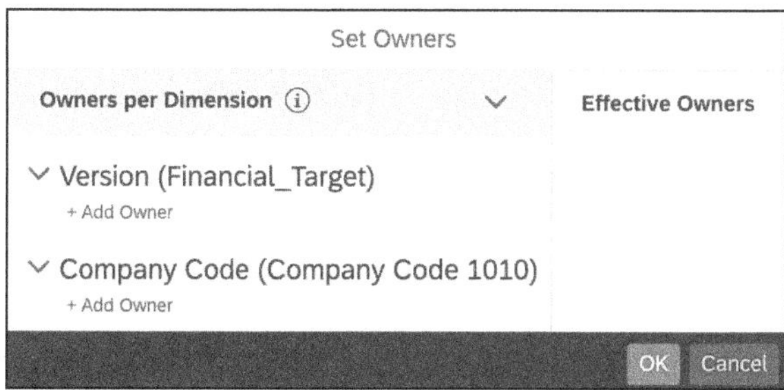

Figure 4.30 Defining Owners per Driving Dimension

You've seen how data locks are set up and managed via the modeler. As mentioned above, this is also possible within a story. To demonstrate this, we've created a table within a story based on our planning model. In the context menu of this table, you'll find the **Manage Data Locks ...** item (see Figure 4.31), and if you click on it, the window for managing the data lock opens (refer back to Figure 4.28). The big difference between this and data lock configuration via the modeler is that you can't assign ownership in this way.

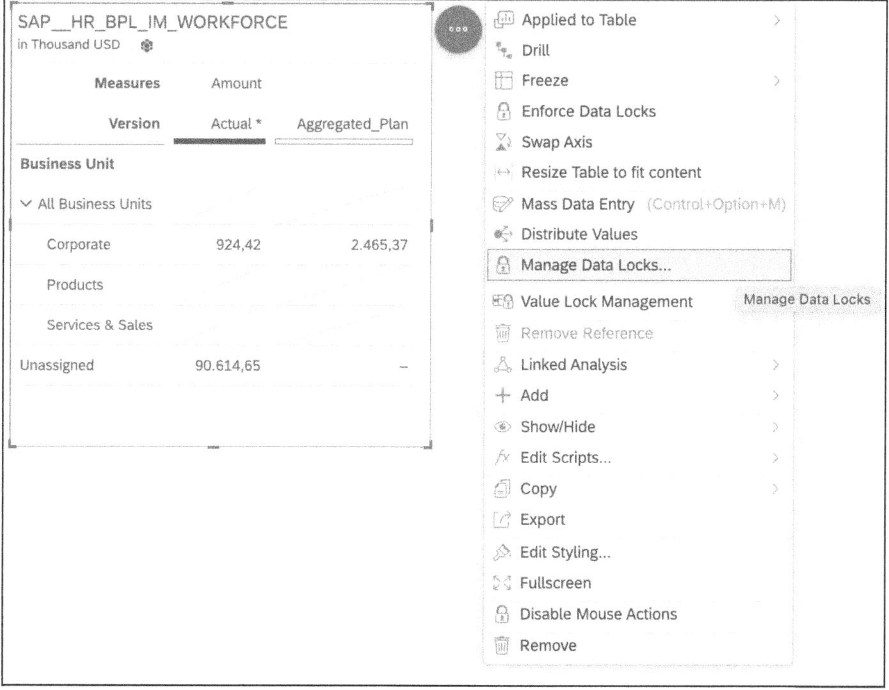

Figure 4.31 Managing Data Locks via Table

The manual method shown here is suitable for smaller planning processes. However, if you are working on a larger planning process in which different data lock states frequently occur, this process can also be a source of errors or can take additional time. Depending on the complexity and importance of the planning process, you should therefore consider whether you can execute such administrative processes in a script-controlled manner.

In addition to setting the data lock manually, you can use the API, which offers you various methods for using the data lock via scripting. We present these methods in the following paragraphs. In the first step, we want to read out the status of the data lock of a planning model on which a table is based, and we use the previously created table for this (see Figure 4.31).

Set data lock script-based

To execute the methods of the DataLocking API, you need an event with which you can start the method. In this example, we use an onClick event of a button (see Figure 4.32), but you can also execute the methods via other events.

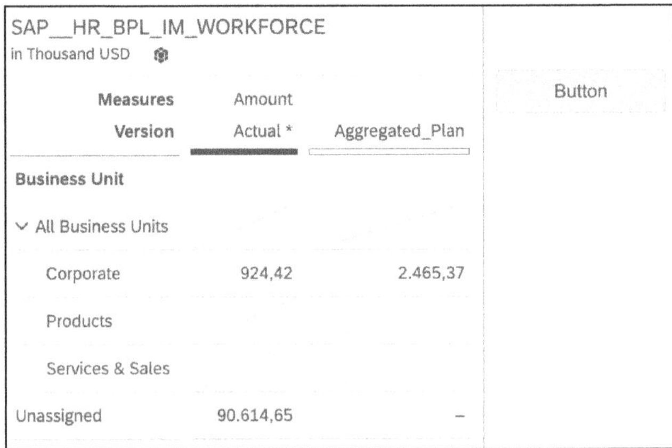

Figure 4.32 Additional Button for Table

Once you've created the button, you can define your script for the onClick event. Since the DataLocking API is part of the planning library, you must first request the planning object of the table, and you can then execute the getDataLocking method in this planning model. You can use this query to check whether a data lock is activated for the planning model, as shown in Listing 4.5.

```
var planning = Table_1.getPlanning();
var dataLocking = planning.getDataLocking();

console.log(dataLocking);
```

Listing 4.5 Checking Data Locking

If the check is successful, you'll receive an object of the `DataLocking` type as the result, and if no data lock is activated for the model, you'll receive the `undefined` value as the result of the query. Because you've previously activated a data lock for your planning model, you can see in Figure 4.33 that an object of the `DataLocking` type is correctly output.

```
Table_1#Planning#DataLocking
```

Figure 4.33 Console Log of DataLocking Object

To query the status of the data lock, you change the structure of your table. Select **Company Code** as the row dimension and filter the table to the year 2025. The table should now look like Figure 4.34.

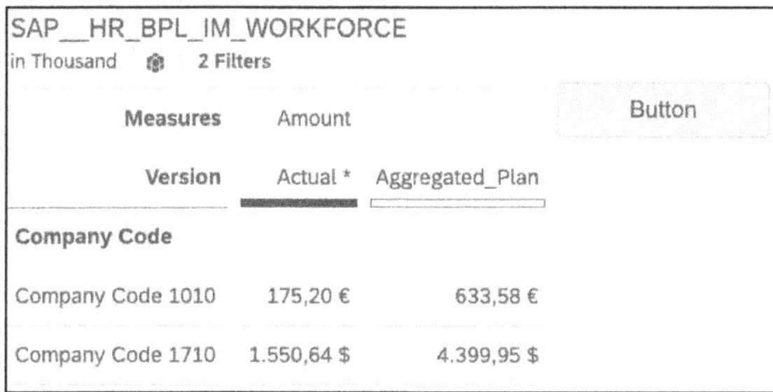

Figure 4.34 New Table Structure

Next, use the data lock function to make changes to the lock status of individual data slices so that you can query them. In this use case, you restrict the **Aggregated_Plan** version for **Company Code 1010** and lock the **Actual** version for **Company Code 1710**. Figure 4.35 shows the final data locking state.

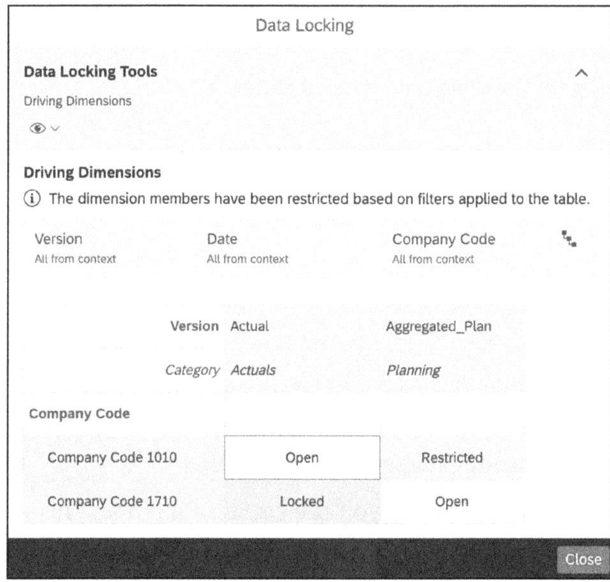

Figure 4.35 Assignment of Data Locking States

After these preparatory steps, you can customize the script further. The DataLocking object offers you two methods for this: getState and setState. The getState method still requires you to select a table cell as an argument because if this argument is not given, the method returns undefined. For this reason, a table query is added to the script (see Listing 4.6).

```
var selection = Table_1.getSelections()[0];
var dataLockingState = Table_
1.getPlanning().getDataLocking().getState(selection);

console.log("data locking state:");
console.log(dataLockingState);
```

Listing 4.6 Script to Log Data Locking State of Selected Cell

In the first line of Listing 4.6, the script queries which table cell is currently selected. In the second line, this variable is used as input for the getState method. The lock state is then output via the console. To check the correctness of your scripts, select the data cells in the table from left to right and from top to bottom, and after each selection, press the button that you used to implement the lock state query. Your result should look like Figure 4.36 and match the definition in Figure 4.35.

```
data locking state:
Open
data locking state:
Restricted
data locking state:
Locked
data locking state:
Open
```

Figure 4.36 Console Log for Checking Data Locking State

Return values of the getState method

The getState method can return a total of four different states:

- Open
- Mixed
- Restricted
- Locked

The Mixed state indicates that the data lock status can't be determined. There can be several reasons for this; for example, the elements in the selection may have different data lock states.

In the next script example, we'll show you the setState method. This requires the selection (i.e., the table cell whose lock state is to be changed) and the new state as a DataLockingState object as arguments. The DataLocking-State objects that you can use are as follows:

- DataLockingState.Open
- DataLockingState.Restricted
- DataLockingState.Locked

Listing 4.7 redefines the onClick event of a button. It reads the current lock status of the table cell and outputs it via the console, and it then locks the selected cell and issues a message as to whether this process was successful. The new lock status is then communicated again via the console.

```
var selection = Table_1.getSelections()[0];
var oldDataLockingState = Table_
1.getPlanning().getDataLocking().getState(selection);
console.log("old data locking state:");
console.log(oldDataLockingState);

var isSet = Table_
1.getPlanning().getDataLocking().setState(selection,
DataLockingState.Locked);
console.log("data locking successfull:");
console.log(isSet);

var newDataLockingState = Table_
1.getPlanning().getDataLocking().getState(selection);
console.log("new data locking state:");
console.log(newDataLockingState);
```

Listing 4.7 Setting Data Locks

Select the cell at the top left with the **Company Code 1010** dimension element as the **Company Code** and **Actual** as the **Version** (refer back to Figure 4.35), then execute the script by clicking the button you used to implement the change to the lock status. Your console output should look like Figure 4.37, which will mean the lock state change has been successfully implemented. The selection was initially **Open** and is now **Locked**.

```
old data locking state:
Open
data locking successfull:
true
new data locking state:
Locked
```

Figure 4.37 Data Lock State via Console Logging

You can also perform the same check again via data lock management. Open this directly via the context menu of the table. Figure 4.38 shows you the change.

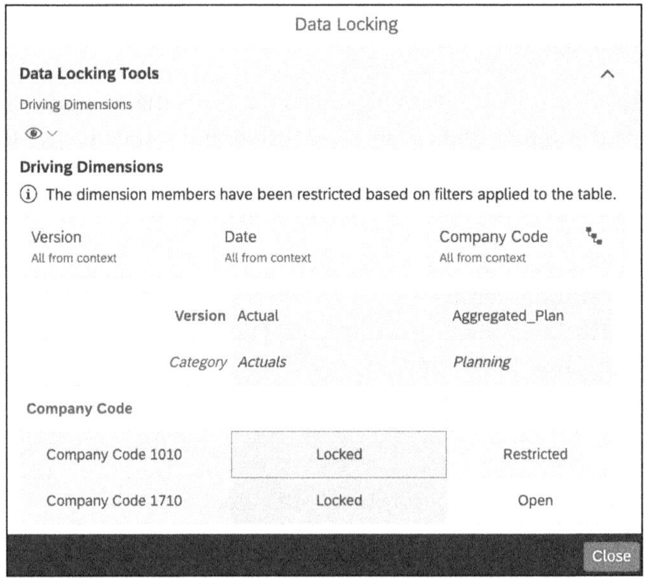

Figure 4.38 Data Locking State after Script-Based Data Lock Changes

MultiActionTrigger

You can use the `MultiActionTrigger` API to extend the multi action trigger widget, and you can use the `onBeforeExecute` event to read whether a multi action has been executed. The event behaves in the same way for multi actions as it does for data actions and planning sequences, and the event is executed when the multi action trigger widget is clicked. If the event returns `false`, the multi action has not been executed, and if the value is `true` or no value, the multi action has been executed.

Planning

Methods for tables and data sources

The `Planning` API offers many methods in terms of tables and data sources:

- **`getDataLocking`**
  This method returns the `DataLocking` object of a table. If the model that is stored in the table doesn't support data locking, the `undefined` value is returned.
- **`getPlanningAreaInfo`**
  Use this method to read the planning area info object that is initialized with the filters associated with the table. If the data source associated with the table doesn't support planning, the return value is `undefined`.

- `getPrivateVersion`
  Use this method to request a specific private version of a model using an ID. If the model doesn't have a private version associated with the specified ID, `undefined` is returned.
- `getPrivateVersions`
  The results of this method are all private versions of a data model.
- `getPublicVersion`
  This method behaves in the same way as the `getPrivateVersion` method, with the difference that the `getPublicVersion` method can be applied to public versions.
- `getPublicVersions`
  This method behaves in the same way as the `getPrivateVersions` method, with the difference that the `getPublicVersions` method can be applied to public versions.
- `isEnabled`
  This method can be used on tables, and it returns `true` if a table has planning enabled and `false` if it doesn't.
- `setEnabled`
  This method activates and deactivates the planning property of a table. If this method is successful, you receive `true` as the result value; if not, you receive `false`.
- `setUserInput`
  This method allows you to write values to a data cell in a table. If this method is successful, you'll receive `true` as the result value; if not, you'll receive `false`.

[«]

Additional Information for the setUserInput Method

If the `setUserInput` method is not successful, it may either be because the data cell is a fact cell or a calculated cell or because the cell is not activated for input. Therefore, after the `setUserInput` method has been executed, we highly recommend that you submit the input using the `submitData` method. Otherwise, your input could be discarded by further user actions or script executions.

- `submitData`
  This method transfers the data to the server. If the transfer is successful, the value `true` is returned; otherwise, the value `false` is returned.

4.2.5 PlanningModel

Planning model

Unlike analytic models, there are methods that can only be applied to planning models. You can use the `PlanningModel` API for this, and the following methods are available to you with this API:

- `createMembers`
  This method creates new dimensions in a planning model. If it is successful, `true` is returned, and if unsuccessful, `false` is returned. It's not possible to create multiple dimensions with the same ID, so if you want to create a new dimension and the ID has already been assigned, the method is cancelled with an error.
- `deleteMembers`
  Similar to the previous method, this method deletes dimensions of a planning model. If successful, `true` is returned, and if unsuccessful, `false` is returned.
- `getMember`
  Use this method to return individual dimensions.
- `getMembers`
  Use this method to display several dimensions of a planning model. For example, you can use optional arguments of the method to set how many elements should be returned in total or whether a fixed number of elements should be skipped at the beginning.
- `updateMembers`
  Use this method to update dimensions. If successful, `true` is returned, and if unsuccessful, `false` is returned.

Limitations of the Methods and Adoption of Changes

The methods we have listed currently only support generic dimensions.

After making changes to the planning model (for example, by creating, deleting, or updating dimensions), you should execute the `DataSource.refreshData()` or `Application.refreshData()` method. This ensures that your changes are also transferred to the subsequent method calls.

4.2.6 DataSourceComments

Comments

The `DataSourceComments` API allows you to add, delete, and edit comments using methods. The following methods are available for this purpose with this API:

- `addComment`
  You can use this method to add one comment to a data cell. If the operation has been executed successfully, the comment is returned; otherwise, `undefined` is returned.
- `addComments`
  Similar to the previous method, this method allows you to add several comments to a data cell. The same applies to the result value. If the operation is successful, the added comments are output, and if the operation fails, the value `undefined` is returned.
- `getAllComments`
  You can use this method to retrieve all comments from a data cell.
- `getComment`
  You can use this method to retrieve an individual comment from a data cell by using the comment ID.
- `removeAllComments`
  You can use this method to remove all comments from a data cell. After completing this method, you receive a Boolean value indicating whether the method was successful or not.
- `removeComment`
  You can use this method to remove an individual comment. To do this, you need the comment ID. After executing the method, you receive a Boolean as the result value, as with the previous method.
- `setCommentLiked`
  You can use this method to define whether a comment should be liked. To do this, you need the comment ID and a Boolean that defines the status.
- `updateComment`
  You can use this method to update a comment. To do this, you need the comment ID and a Boolean that defines the status.

[!]

Updating Comments

The `updateComment` method is only supported for SAP BW live data models.

[!]

Mobile Support

The methods of the `DataSourceComments` API can't be used on mobile devices.

4.3 Data Connection for Writing Back Data

Open Data Protocol

SAP Analytics Cloud uses the *Open Data Protocol* (*OData*) to communicate with data sources and to write back plan data. OData is an HTTP-based protocol published by Microsoft that enables data exchange between software systems. Due to its standardized semantics, the use of OData is particularly popular in the area of cloud solutions, as this protocol can be used to implement client-server communication.

Important terms

In the context of OData services, there are some terms that are not familiar to everyone who hasn't worked with OData services before. We therefore explain some of these terms in more detail as follows:

- *OData actions* are operations that are offered by an OData service and whose call can cause side effects. Actions can return data but may not be further combined with additional path segments.
- *OData import actions* or *unbound actions* are not linked to an entity set or an entity. They generally provide simple parameters, and all parameters are set via the POST body.
- *Bound actions* are actions that can be called for a specific entity. They always contain a first parameter that is set via the URL (to resolve the binding to the entity in the corresponding entity set), and all other parameters are set via the POST body. In general, actions can be bound to any type, but in stories, only the binding to individual entities is supported. OData actions can be called and executed from the backend system by executing program script within a story. Programmatic read access to OData services is also provided.

In the following sections, we'll give a brief introduction to OData in SAP Analytics Cloud. After that, we'll show you how to set up an OData connection and how to use it to read data.

4.3.1 OData in Stories

OData services can be defined in stories based on existing live connections. These connections must be of the type *cross-origin resource sharing* (*CORS*) type or the *direct connections* type. You can use OData services to define connections with the following systems:

- SAP BPC
- SAP BW
- SAP HANA
- SAP S/4HANA

[«]

OData Configuration

Cross-origin resource sharing should be configured for OData in the same way as an information access connection. You should also define `if-match` as a permitted header.

Characteristics of the OData connection

If you decide to execute transactional processes in the backend system via OData calls, for example, you should always observe the following points:

- OData actions are processes of an OData service, and their execution can have side effects. OData import actions (also known as unbound actions) are not linked to OData entity sets or entities. Import actions have simple parameters, and all parameters are defined via a POST body.
- Actions that are called for a specific entity are called bound actions. They always define a first parameter via URL, and this is used to resolve the binding to the entity within the relevant entity set. As with unbound actions, the remaining parameters are defined via the POST body.
- OData actions can be bound to any type. In stories, however, the binding is restricted to individual entities.

These points were taken from the overview of the use of OData services in "Optimized Story Experience" in the SAP Help Portal, which you can access at *http://s-prs.co/v594403*. You'll always find up-to-date information on this topic on this web page.

When you're using the OData service, you should be aware of some restrictions to familiarize yourself with before creating your story. These restrictions include the fact that only simple parameter types are supported. Actions that contain mandatory parameters of unsupported types are not available in analytical applications, and for actions with optional parameters of unsupported types, the action itself is supported but the parameters are not.

For bound actions, actions that are bound to entity sets are not supported. Only actions that are linked to a single entity are supported, and you can only specify binding parameters if you specify a key to the entity for which the action is to be executed. Please also note that only the JSON format is supported and that OData connections can only be used in the on-premise version of SAP S/4HANA.

The following types are not supported:

- `Edm.Stream`
- `Edm.Untyped`
- All `Edm.Geography` types
- All types defined in another namespace

These restrictions were taken from the overview of the use of the OData service in stories in the SAP Help Portal, which you can access at *http://s-prs.co/v594403*.

4.3.2 Setting Up OData Services

Set up OData connection

To create and set up new OData services in your story, you must open the **Outline**, which opens as a panel on the left-hand side of the screen. As you can see in Figure 4.39, the **Outline** is divided into two areas: **Story** and **Scripting**. A story can contain many different artifacts that can be used for scripting. To create a new artifact, click the **+** icon.

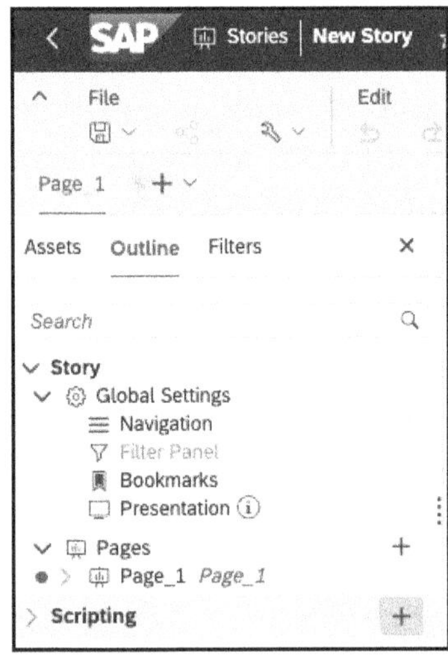

Figure 4.39 Outline View of Left-Side Panel

In the popup window shown in Figure 4.40, click on **OData Services**.

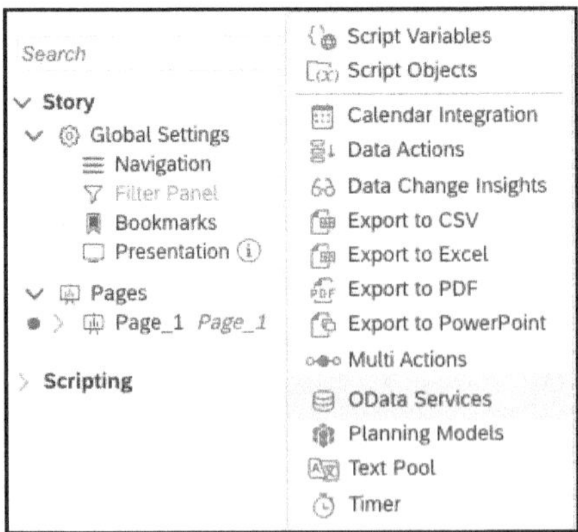

Figure 4.40 Popup to Create New Scripting Artifacts

Once you've created a new OData service, it's listed in the **OData Services** area. The newly created OData service is given a generic name by the system, and in Figure 4.41, it is named **ODataService_1**. You have the option of changing the name after creation according to your wishes.

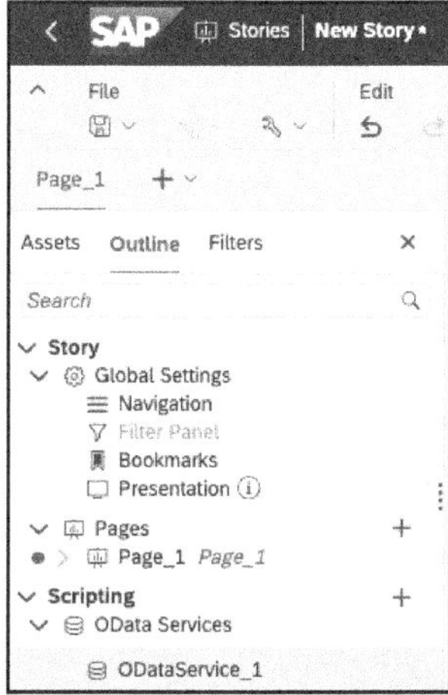

Figure 4.41 New OData Service Added to Story

After you create the service, you need to define its properties because it's still an empty object that can't make any calls. To do this, select the corresponding OData service in the outline. A panel should then open on the right-hand side of your screen (see Figure 4.42), and you can use the panel to maintain the properties of the OData service. The panel allows you to define three properties:

- **Name**
- **Connection**
- **End-Point URL**

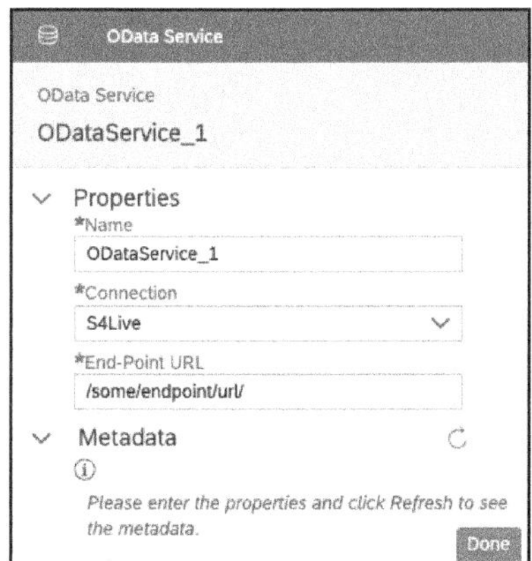

Figure 4.42 Properties Definition of OData Service

The **Name** field allows you to change the default name assigned by the system and assign a meaningful name.

For the **Connection** property, you want to select an existing connection to a source system from a dropdown list, so you need to have already created a connection to the source system before you use the OData service. Note the possible connection types that we've defined in Section 4.3.1.

You must specify the **End-Point URL** as the third property. Each OData service has its own URL, and you must specify a valid URL here.

[»]

No Input Assistance for Endpoint URL

Unfortunately, there's no input assistance when you're entering the endpoint URL after selecting the connection. There's also no catalog or dropdown list that you can use to select a valid endpoint URL, so you must already know the URL in advance.

If you've entered a connection with a valid URL, you can call up the *metadata* (data that provides information about other data—for example, actions that are listed) of the OData service in the **Metadata** area. To do this, you must first update the metadata.

4.3.3 Reading Data from an OData Service

Using an OData connection

With OData services in stories, you can not only call and execute OData actions, but you can also read out the data provided. In this way, you can use the information provided via entity sets (for example, if you want to display individual entities in a text widget).

The exact use of the APIs for reading OData services is described in Section 4.2.2.

[!]

Restrictions when Reading Data

You must take the following restrictions into account when reading data provided via OData services:

- The *concatenation* (chaining) of multiple entity sets is not supported.
- The EXPAND option (analogous to JOIN) is not supported.
- Entity sets with parameters are not supported.
- Entity sets with mandatory filters are not supported.

These restrictions were taken from the SAP Help Portal overview of the use of OData service in stories, which you can access at *http://s-prs.co/v594403*. On this page, you'll always find the current status of the restrictions as well as further information.

4.4 Developing a Story with Scripting for Planning

In this section, we'll show you how to use some of the methods presented in this chapter to create a planning story with scripting functionality. The aim of the story is to provide end users with an interface that they can use in their planning process. The story should allow users to change existing plan values and create new dimension elements, and collaborative functions such as leaving comments should also be possible in the story.

> **Simplicity of the Example**
>
> We've deliberately kept the story in this section simple. It should be easy for you to develop parts of the example yourself, and in Section 4.5, we also look at examples that require more development effort.

Structure of this section

The central element of the planning application is a table, and you'll develop the story in separate steps. In the first step, you'll add functions based on the `Planning` API. In this section, the user should be able to change plan values in the table via an input screen. In Section 4.4.2, "Use Cases of Specific Methods of the Planning Application Programming Interface," you'll use the `PlanningModel` API to extend the functions of the planning application, which enables users to add new dimension attributes. In Section 4.4.3, "Examples of Specific Methods of the PlanningModel Application Programming Interface," you'll learn how to extend the application to include collaborative elements. By using the `DataSourceComments` API, you'll give users the opportunity to create comments and annotations for colleagues and exchange information with them.

Figure 4.43 shows the final story. In addition to the functionalities already mentioned, the application is supplemented by a filter element.

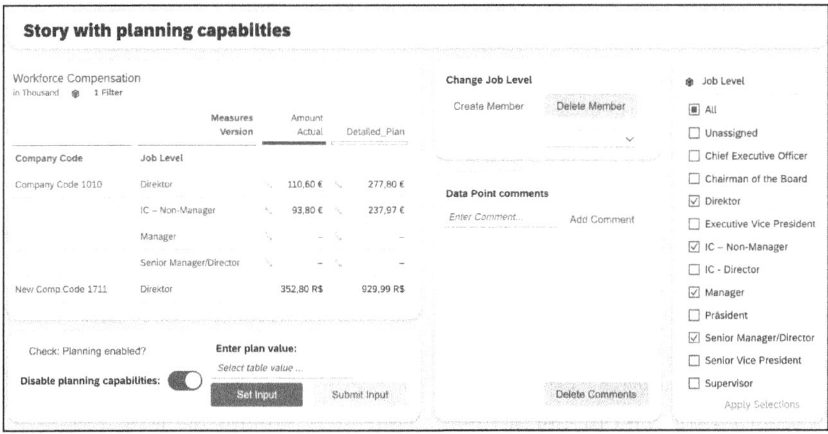

Figure 4.43 Story to Demonstrate Specific Planning Methods

4.4.1 Preparation of the Sample Story

Before we take a closer look at individual use cases in the next sections, we'll present a method for improving the user experience for end users in this section.

When you open the story, individual widgets are loaded at different speeds. Especially on different end devices, there may be variations in overall loading time, depending on system performance. It can be very annoying for end users if an object in the story is already loaded but they can't interact with it or the story doesn't respond to their interaction because there are dependencies in the background to other objects in the application that are not yet fully available.

onInitialization event

For this reason, you need to display a message at the start of the story that the story is still loading. To do this, you use the `onInitialization` event, which is executed when the story is started. The notification that you see in Figure 4.44 is created by the `showBusyIndicator` method.

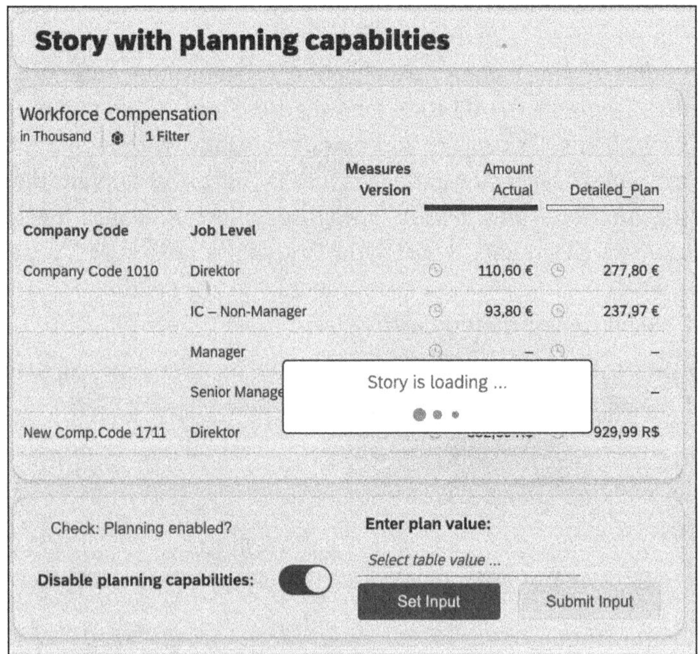

Figure 4.44 Loading Screen when Opening Story

As the indicator must be hidden again with the `hideBusyIndicator` method, you use a loop that delays the hiding of the indicator until the sample application is ready (see Listing 4.8). The loop is a `while` statement that tests whether the data model of the table already contains data. As long as no data is available, the query is repeated, but as soon as data is available, the script is continued and the message is hidden.

```
Application.showBusyIndicator('Story is loading ...');
while (Table_1.getDataSource().isResultEmpty()) {
     console.log("Story is loading ...");
}
ScriptObject_1.updateDropdown();

Application.hideBusyIndicator();
```

Listing 4.8 onInitialization Event of Page

At the end of the script, a script function of the `ScriptObject_1` script object is executed. The functionality and meaning of this function are explained in Section 4.4.3.

4.4.2 Use Cases of Specific Methods of the Planning Application Programming Interface

In this section, we look at the first part of the example story. The main focus here is on the methods of the `planning` API, and Figure 4.45 shows the structure of the planning story.

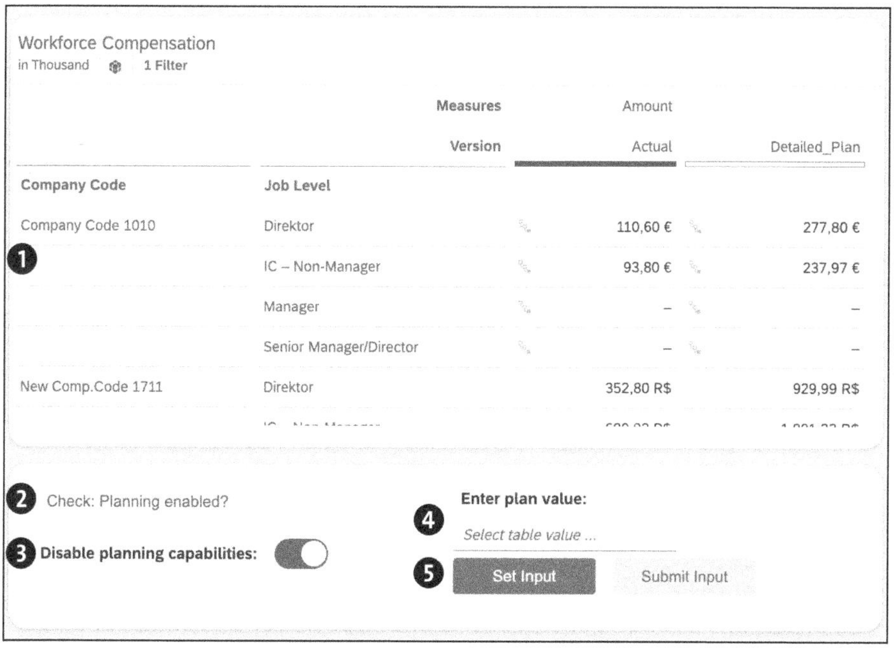

Figure 4.45 Part of Story Demonstrating Planning API

Functional scope of the example

You can use this example to test various methods in a planning context. The central element of this story is a table, and you can work with this table using various input methods to carry out the following process steps:

- You can check whether the table's planning capability is enabled.
- You can enable or disable the planning property of the table.
- You can enter new plan values using the text input.
- You can submit plan values to the planning model and write them back.

Structure of this part of the story

Before we go into the specific methods and script fragments in more detail, we'll explain the structure of the story, which is divided into five areas, as shown in Figure 4.45:

❶ The table was created with the `SAP__HR_BPL_IM_WORKFORCE` planning model, and the **Company Code** dimension was added as a row dimension. At row level, **Version** is selected as a dimension and filtered to the **Actual**, **Aggregated_Plan**, and **Detailed_Plan** attributes.

❷ This element is a button, and the text has been adapted to improve usability.

❸ This area is a combination of a text widget (to describe the functionality) and a switch widget. The state of the switch is set to "On" by default.

❹ The user input of the plan value is made possible by an input field. In the context of the story, we have added a note that the user must select a table cell.

❺ This area consists of two buttons, and we've adapted the text to describe the function behind it. Changing the type of button style has no functional relevance but is intended to simplify the visual differentiation.

Checking the Planning Capabilities

Test planning capabilities

Checking whether a table has enabled the planning capabilities is quite simple with the `isEnabled` method, and you can apply this method directly to the table widget. In this example, we output the result of the check via a message, and to start the script, the story uses the `onClick` event of a button (see Listing 4.9).

```
if (Table_1.isEnabled()) {
    Application.showMessage(ApplicationMessageType.Success, "Planning 
capabilities are enabled.");
} else {
    Application.showMessage(ApplicationMessageType.Warning, "Planning 
capabilities are disabled.");
}
```

Listing 4.9 onClick Event of Button

Figure 4.46 shows you the result if you haven't made any further adjustments.

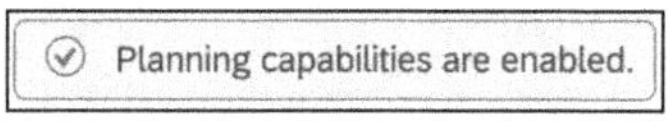

Planning capabilities are enabled.

Figure 4.46 Message from Planning Capability Check

Enable/Disable Planning Capabilities

Set planning capabilities

Now that you know how to check a table for its planning capability, let's look at how you can set the table's planning capability. You can do this with the `setEnabled` method, which requires a Boolean, the future state of the planning capability, as input. You obtain this Boolean from the switch widget.

The first two script lines in Listing 4.10 describe this procedure. The switch state is saved in the `switchState` variable, and it is then used as an argument for the `setEnabled` method.

The user is then given feedback with an `if` statement. The descriptive text of the text field is adjusted accordingly, and a message is displayed to indicate whether the planning function has been enabled or disabled.

```
var switchState = Switch_1.isOn();

Table_1.setEnabled(switchState);

if (switchState) {
     Text_2.applyText("Disable planning capabilities:");
     Application.showMessage(ApplicationMessageType.Info, "Planning
capabilities enabled.");
} else {
     Text_2.applyText("Enable planning capabilities:");
     Application.showMessage(ApplicationMessageType.Info, "Planning
capabilities disabled.");
};
```

Listing 4.10 Script of Switch's onChange Event

You can see the use of the script from Listing 4.10 in Figure 4.47. Here, the planning capabilities have been disabled and the switch has been turned off. Accordingly, as expected, the system message has been displayed and the text in the text field has changed.

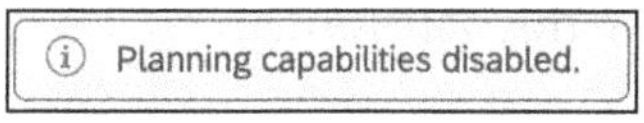

Planning capabilities disabled.

Figure 4.47 Message Created Using Switch to Disable Planning Capabilities

If you now click on the button to check the planning capability again, the message should display the desired result. Figure 4.48 shows you the message.

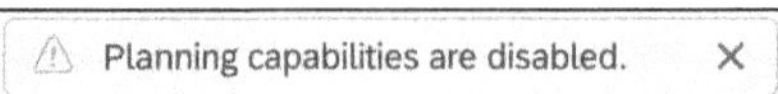

Figure 4.48 Message from Planning Capability Check after Disabling Planning Capabilities

Submission of the Input

Writing back user input

In the next part of the story, you can edit the planning values of the model. You use the `setUserInput` method for this, and as a user, you have the option of entering new numerical values via an input field. You then start the event that calls the method by clicking the **Set Input** button.

As you can see from the script, the `setUserInput` method is not executed directly. Checks are carried out beforehand to ensure that the input is correct, and then, three more checks are carried out as follows:

1. If you want to modify a plan value of a table cell, you must first select a table cell. The first `if` statement checks whether a table cell in the table has been selected, and if it hasn't, a message will be displayed and the event will be terminated.
2. If you want to modify a plan value of a table cell, you must enter a plan value. The second `if` statement checks whether a plan value has been entered via the input field, and if it hasn't, a message will be displayed and the event will be ended.
3. You can only use integers or decimal numbers when entering planning data. This check is automated by the system, so you don't have to carry out a check yourself. If you want to write back a string using the `setUserInput` method, the action will be aborted, and a corresponding message will be displayed.

Implementing checkSelection as script function

Because the check of the selected table cell must be done frequently when you're using the story, it's recommended that you implement this as a separate function. You can use *script objects* for this purpose. Figure 4.49 shows you the definition of the `checkSelection` script function, and there are two settings to note here. `Boolean` has been set as the return type, which means that the function returns a Boolean—in this case, the `true` value if the check was successful and `false` if the check detected an error. Also, an argument has been defined. The `selections` argument is an array of the `Selection` type, and the selection of table cells is passed in this data format.

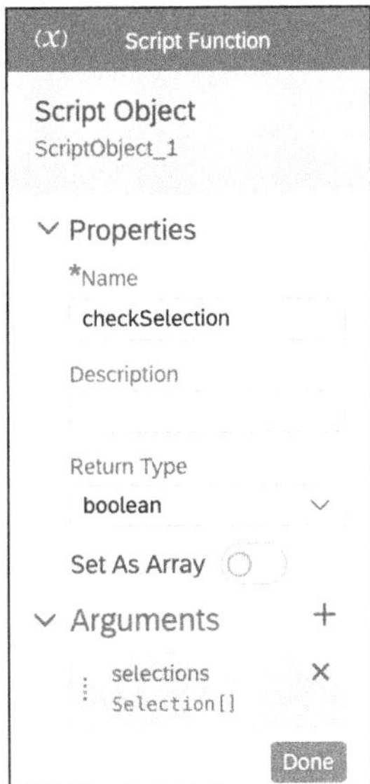

Figure 4.49 Definition of checkSelection Script Function

The script function consists of a sequence of three `if` statements. The function performs different queries:

1. It checks whether the `selections` argument represents table cells.
2. It checks whether the argument represents only one and not several table cells.
3. It checks whether the argument represents a data cell.

How to use script variables

If one of these queries is negative, the function is aborted and `false` is returned. In addition, the input field of the new plan value and the comment field (which we discuss in Section 4.4.4) are reset. If the three `if` statements are successfully executed, the `data` and `selection` *script variables* are defined before the script is completed. The `data` script variable is an object of the `DataCell` type, and after the selection has been successfully checked, the data value of the table cell is saved in this variable so that it can be called by other methods within the story. The `selection` script variable is an object of the `Selection` type, and once the selection has been successfully checked, it's saved in the script variable. You can see this process in Listing 4.11.

```
if (selections === undefined) {
    Application.showMessage(ApplicationMessageType.Warning, "Please
select a table cell");
    return false;
}

if (selections.length > 1) {
```

```
    Application.showMessage(ApplicationMessageType.Warning, "Please
select only one table cell");
    InputField_NewValue.setValue("");
    Text_Comments.applyText(" ");
    return false;
}

var data_$ = Table_1.getDataSource().getData(selections[0]);

if (data_$ === undefined){
    Application.showMessage(ApplicationMessageType.Warning, "Please
select a data cell");
    InputField_NewValue.setValue("");
    Text_Comments.applyText(" ");
    return false;
}

selection = selections[0];
data = data_$;

return true;
```

Listing 4.11 checkSelection Script Function

This also means that this information can be called up by other methods of the story via the script variable. You can see the definitions of both variables in Figure 4.50 and Figure 4.51.

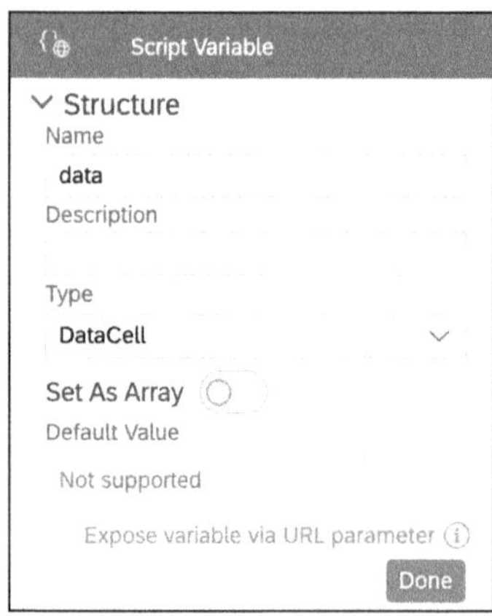

Figure 4.50 Definition of Data Script Variable

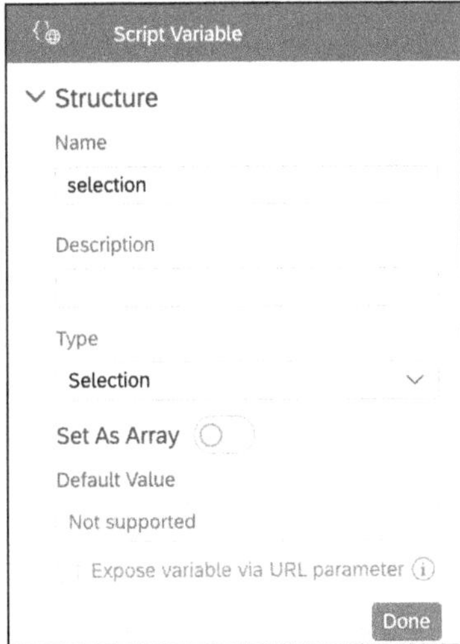

Figure 4.51 Definition of Selection Script Variable

How the script works

If the script is not aborted by the `if` statements, then the `setUserInput` method is used. As this method uses a Boolean as the result value to indicate whether the execution was successful, this value is collected using the `setInput` variable. A case decision is then used to provide feedback as to whether the operation was successfully executed or aborted.

Figure 4.52 shows you the sample story with the warning messages generated by the script.

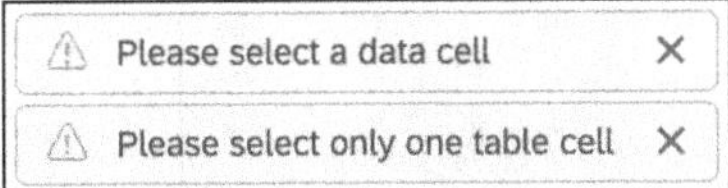

Figure 4.52 Warning Messages Created by checkSelection Script Function

Simplifying the Story with Table Functions

How to use table functions

To make it easier for users to enter the plan values, you can use a table event. When selecting the table cell, the table widget offers the option of using an `onSelect` event. In this example, the script performs two operations:

1. It checks the table selection.
2. It copies the current plan value of the selected table cell to the input field.

Let's first look at why it makes sense to check the selection of the table cell. The table, as you can see in Figure 4.53, not only consists of plannable table cells but also of descriptive row and series elements that reflect the dimension attributes. These cells can't be used by the `setUserInput` method; it would result in the operation being aborted. For this reason, the `checkSelection` script function of `ScriptObject_1` is used here (see Listing 4.12).

```
var selections = Table_1.getSelections();
var resultCheck = ScriptObject_1.checkSelection(selections);

if (!resultCheck) {
    return;
}

InputField_NewValue.setValue(data.rawValue);
ScriptObject_1.updateComments(selection);
```

Listing 4.12 onSelect Event of the Table_1 Table

How the script works

If the `checkSelection` function doesn't detect any errors, the current table value is transferred to the input field for the user via the script. The unformatted numerical value is transferred if you select the `rawValue` property, and this can be useful if scaling of the display is configured in the table. You can also see such scaling in Figure 4.53, where all table values are displayed in thousands. When the selected value is transferred, the number is displayed *unscaled*, which means that digits that are not displayed due to rounding during scaling are also displayed. You can also see this behavior in Figure 4.53.

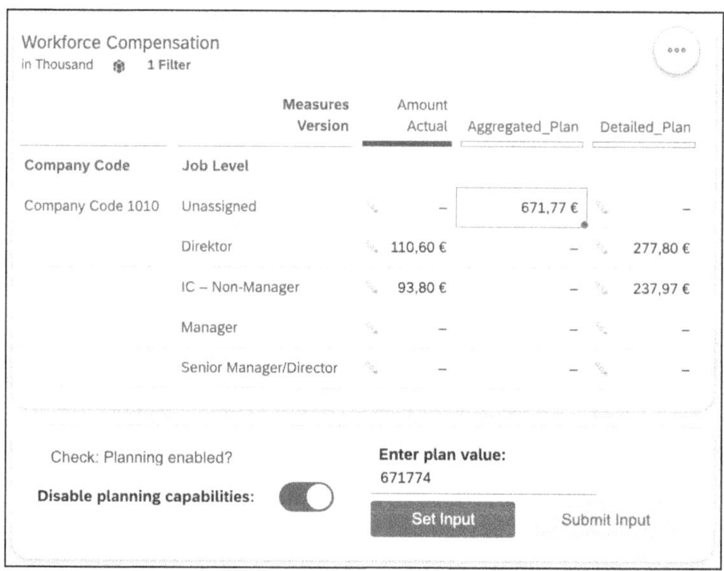

Figure 4.53 Selecting Data Cell

You can use these two functions of the `onSelect` event to simplify the planning process for the user of the planning story.

Submission of the Plan Value

Submit plan values

After the updated plan value has been set by the `setUserInput` method, you can use the `submitData` method to transfer all updated plan values to the server. As all data has already been checked for correctness in advance, you can execute the method directly (see Listing 4.13).

```
InputField_NewValue.setValue("");

var submit = Table_1.getPlanning().submitData();

if (submit) {
    Application.showMessage(ApplicationMessageType.Success, "The
plan value was successfully submitted.");
} else {
    Application.showMessage(ApplicationMessageType.Error, "An error
has occurred during submission.");
}
```

Listing 4.13 onClick Event of the Button_SubmitInput Button

How the script works

The `submitData` method also returns a Boolean as the result value, which in turn indicates whether the operation has been performed successfully. This value is collected by the `submit` variable in this script, and based on this, the user receives feedback via a message as to whether the operation was successful or not.

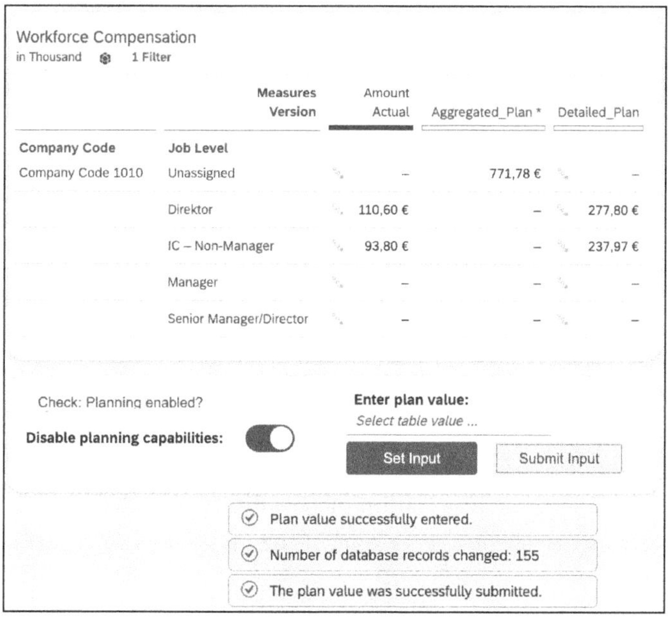

Workforce Compensation
in Thousand 1 Filter

	Measures / Version	Amount / Actual	Aggregated_Plan *	Detailed_Plan
Company Code	**Job Level**			
Company Code 1010	Unassigned	–	771,78 €	–
	Direktor	110,60 €	–	277,80 €
	IC – Non-Manager	93,80 €	–	237,97 €
	Manager	–	–	–
	Senior Manager/Director	–	–	–

Check: Planning enabled?
Disable planning capabilities:
Enter plan value:
Select table value ...
Set Input
Submit Input
Plan value successfully entered.
Number of database records changed: 155
The plan value was successfully submitted.

Figure 4.54 Successful Submission of Planning Data

Figure 4.54 shows the message saying that the `submitData` method has been executed successfully, and it also shows two other messages: the scripted message that is displayed after the `setUserInput` method has been successfully executed and a system-side message. The number of data value changes is displayed at a granular level by transmitting the plan value to the server. In this example, 155 data records have been updated.

4.4.3 Examples of Specific Methods of the PlanningModel Application Programming Interface

In this section, we look at the second part of the example story, and we show you potential application scenarios for the `PlanningModel` API. Figure 4.55 shows the structure of the planning story in this section.

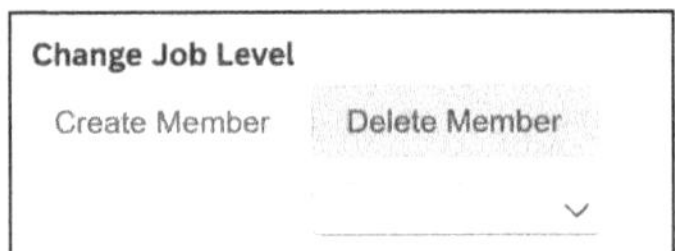

Figure 4.55 Second Section of Story

How the example works

With this example, you'll be able to use two essential functions within a planning process:

- You can add new elements to the **Job Level** dimension, and you can then plan these new elements via the table.
- You can delete existing dimension elements in the **Job Level** dimension.

Structure of the section

Before we go into more detail about how the methods work, let's look at the structure of this part of the story. This part consists of three areas or widgets:

- A button that you use to start the methods for creating a new dimension element
- A button that you use to initiate the function for deleting selected dimension elements
- A dropdown list that you use to list all current dimension elements of the **Job Level** dimension of the planning model

Creating a New Dimension Element

Create dimension member

In this section, we'll look at the creation of new dimension elements. To start the process, we use the `onClick` event of the **Create Member** button (see Listing 4.14).

```
Popup_1.open();

var members = PlanningModel_1.getMembers("SAP_HR_USER_JOBLEVEL");
var text = "";

for (var i = 0; i < members.length; i++) {
     text = text + members[i].id + "\n";
}

Text_AllIDs.applyText(text);
```

Listing 4.14 onClick Event of Button_CreateMember Button

The event doesn't directly create a new element but is used for two other tasks. Firstly, it opens a popup, and secondly, it lists all dimension elements of the **Job Level** dimension and makes them available to the user in the popup. It does this by using a `for` loop; and in an iterative process, the IDs of the dimension elements are saved in a character string and then written to a text field within the popup.

To add the dimension elements to the correct model, the script uses the `PlanningModel_1` scripting artifact, which was defined in advance via the left-side panel (see Figure 4.40). The **PlanningModel_1** object references the **SAP__HR_BPL_IM_WORKFORCE** model (see Figure 4.56).

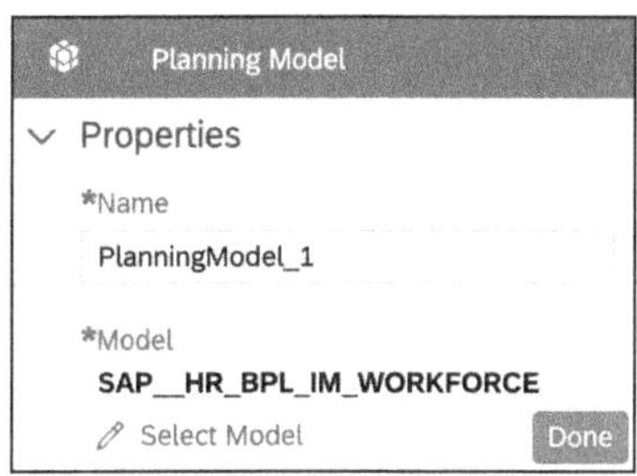

Figure 4.56 Definition of PlanningModel_1

Explanation of the popup

The popup that this script is opens is shown in Figure 4.57. It serves as an input mask for two input values, which is why it has a very simple structure. On the left-hand side, the popup contains two input fields that users can use to enter an element ID and an element description.

As the element ID must be uniquely assigned, all element IDs that have already been assigned are listed on the right-hand side of the popup. The popup also displays descriptive text elements, and there are two buttons in the footer that you have to add via the **Builder** in the right-side panel. Figure 4.58 shows you how to do this.

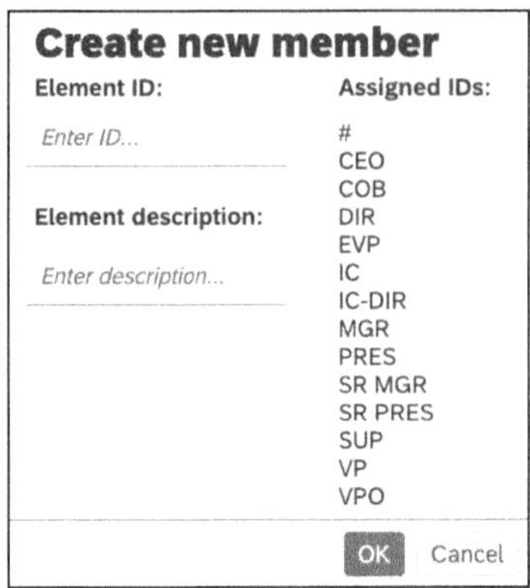

Figure 4.57 Popup Window for Creating New Dimension Members

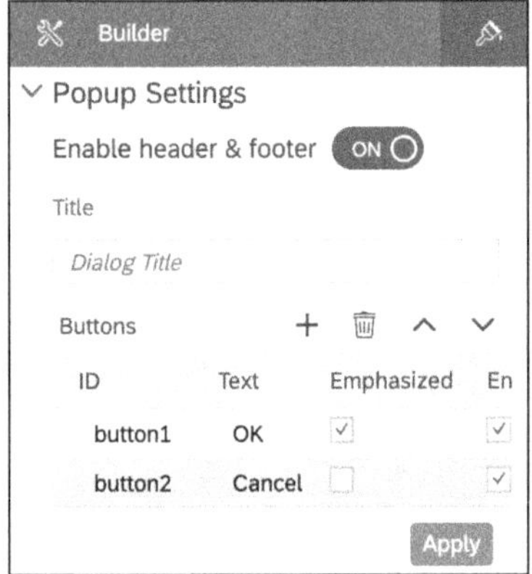

Figure 4.58 Adding Footer to Popup

Define button IDs

Furthermore, you can see here which IDs to use for the respective buttons. You need these to differentiate the user interactions. By adding the footer, you can define a new event for the popup: the `onButtonClick` event, which is always executed as soon as a user clicks one of the buttons. As an argument, this event receives a string that reflects the ID of the button. In this example, we use a `switch` statement to differentiate between the IDs, as shown in Listing 4.15.

```
switch(buttonId) {
    case "button1":
        var members = PlanningModel_1.getMembers("SAP_HR_USER_
JOBLEVEL");
        var inputID = InputField_ID.getValue();
        var inputBes = InputField_Description.getValue();

        if (inputID === "") {
            Application.showMessage(ApplicationMessageType.Error,
```

```
"Please enter an ID.");
                return;
            }

            if (inputBes === "") {
                Application.showMessage(ApplicationMessageType.Error,
"Please enter a description.");
                return;
            }

            for (var i = 0; i < members.length; i++) {
                if (inputID === members[i].id) {

Application.showMessage(ApplicationMessageType.Error, "ID is already
taken. Please enter another ID.");
                    return;
                }
            }

            PlanningModel_1.createMembers("SAP_HR_USER_JOBLEVEL", {
                id: inputID,
                description: inputBes
            });

            InputField_ID.setValue("");
            InputField_Description.setValue("");
            Application.refreshData();
            ScriptObject_1.updateDropdown();
            Popup_1.close();
            Application.showMessage(ApplicationMessageType.Success,
"New Job Level \""+ inputBes + "\" created.");
            return;

        case "button2":
            Popup_1.close();
            return;
}
```

Listing 4.15 Differentiation of Buttons in Footer of Popup

How the script works

The script starts directly with the switch statement. If the second button is clicked, the popup is closed and the function is ended, but if the first button is clicked, checks are carried out first. Various if statements are used to check whether the user has entered an ID and a description in the corresponding input fields, and the system also checks whether the ID has already been assigned. If one of these three checks is positive, a corresponding error message is displayed. Figure 4.59 shows a scenario in which each error message is displayed in turn.

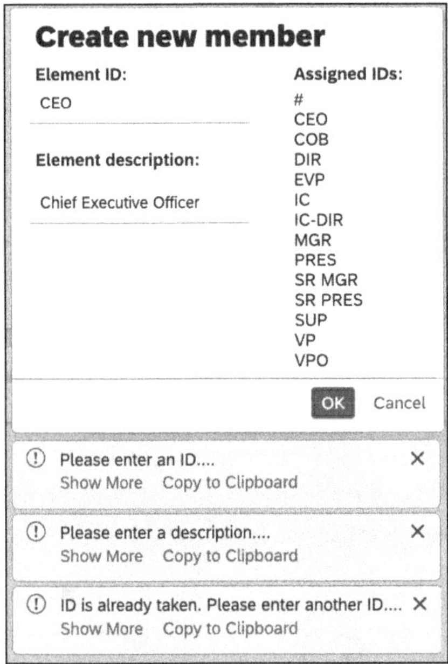

Figure 4.59 Error Messages when Creating New Dimension Member

If no error message is output and the script continues, the `createMembers` method is executed in the next step. With this method, you add a new element to the planning model.

Finally, the input fields of the popup are reset via the script and the story data is updated with the `refreshData` call. This call makes the new element directly visible in the table and the filter. Before the popup is closed and a message is displayed for users, another script function is executed (see Figure 4.60).

Figure 4.60 Successful Creation of New Chief Sustainability Officer Dimension Member

How to use the script function updateDropdown

The `updateDropdown` script function updates the element list of the drop-down to select possible dimension elements for deletion. You must also add the newly created element to this list (see Listing 4.16).

No argument is required to call the script function. The function also doesn't return a result, so the result type is void.

```
var members = PlanningModel_1.getMembers("SAP_HR_USER_JOBLEVEL");
Dropdown_JobLevel.removeAllItems();

for (var i = 0; i < members.length; i++) {
    Dropdown_JobLevel.addItem(members[i].id, members[
i].description);
}
```

Listing 4.16 updateDropdown Method as Part of ScriptObject_1

The script function first deletes all elements of the dropdown list and then adds all dimension members of the **Job Level** dimension to the dropdown list via a for loop.

Figure 4.60 shows you the message that is generated when the onButton-Click function has been successfully executed. In this example, the **Chief Sustainability Officer** element has been added to the **Job Level** dimension, and in Figure 4.61, you can see that the newly created role is directly available in the filter.

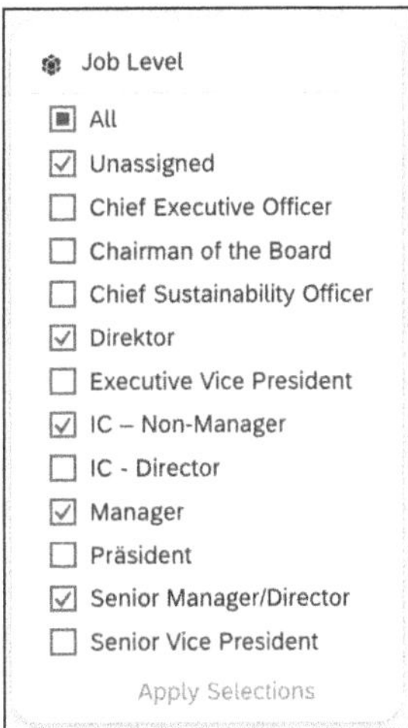

Figure 4.61 New Chief Sustainability Officer Dimension Member as Selectable Element of Filter

If you select **Chief Sustainability Officer** in the filter, a new row appears in the table and you can edit the corresponding data cells using planning functions (see Figure 4.62).

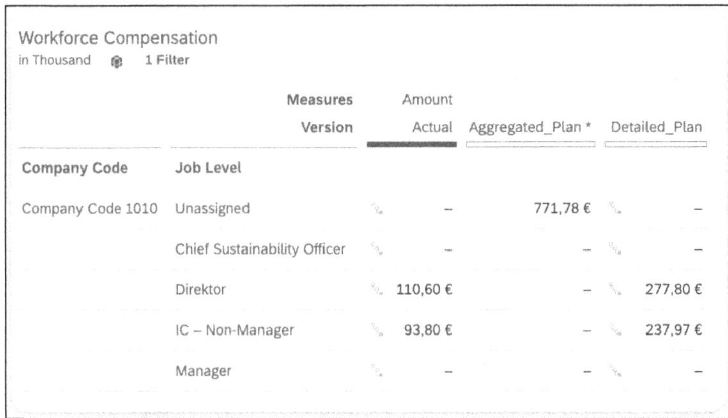

Workforce Compensation
in Thousand 1 Filter

	Measures	Amount		
	Version	Actual	Aggregated_Plan *	Detailed_Plan
Company Code	Job Level			
Company Code 1010	Unassigned	–	771,78 €	–
	Chief Sustainability Officer	–	–	–
	Direktor	110,60 €	–	277,80 €
	IC – Non-Manager	93,80 €	–	237,97 €
	Manager	–	–	–

Figure 4.62 Plannability of New Dimension Member via Table

Deleting Dimension Members

Deletion of dimension member

In this section, we explain how to delete dimension elements using a dropdown widget and a button. The method consists of two steps:

1. Select the dimension element to be deleted from the dropdown list.
2. Click the button to delete the selected element.

How to use the dropdown widget

The first step should be intuitive for users. They can expand the dropdown widget, which will open a vertical list that they can use to select a corresponding element with a single click. Figure 4.63 shows this process.

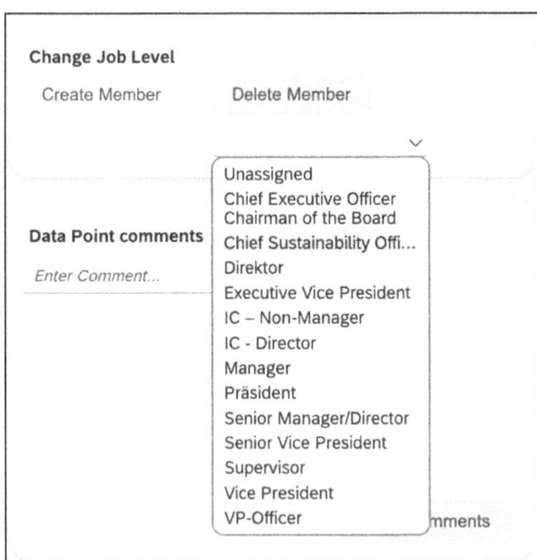

Figure 4.63 Dropdown List for Selecting Dimension Member

As you've seen in the previous section, you can fill the dropdown list with elements by using the `updateDropdown` script function. This function is also executed during the onInitialization event of the story. Listing 4.17 shows you the script to delete a dimension member.

```
var selectedKey = Dropdown_JobLevel.getSelectedKey();
var selectedText = Dropdown_JobLevel.getSelectedText();

if (selectedKey === undefined) {
     Application.showMessage(ApplicationMessageType.Warning, "Please
select an item from the dropdown list.");
     return;
}

PlanningModel_1.deleteMembers("SAP_HR_USER_JOBLEVEL", selectedKey);
Application.refreshData();
ScriptObject_1.updateDropdown();
Application.showMessage(ApplicationMessageType.Info, "Job Level \"" +
selectedText + "\" deleted.");
```

Listing 4.17 Deleting Dimension Member

How the script works

To execute the deletion, use the onClick event of the corresponding button in this story. As part of the script, the system first checks whether an element has been selected via the dropdown list, and if it hasn't, a message is displayed and the function is canceled. Otherwise, the deleteMembers method is applied to the planning model. The selected element from the dropdown list is removed as a dimension element, the story data is updated so that the change is immediately visible, and the dropdown list is updated using the script function. Figure 4.64 shows the message that is displayed to the user after a successful deletion.

ⓘ Job Level "Chief Sustainability Officer" deleted.

Figure 4.64 Confirmation of Deletion of Chief Sustainability Officer Dimension Member

4.4.4 Examples of Individual Methods of the DataSourceComment Application Programming Interface

In this section, we look at the third and final part of the example story, and we show how you can use the methods of the DataSourceComment API. Two functionalities that are particularly interesting for collaborative planning cycles are shown in this part of the example story:

Features of the example

- Creating data point comments for individual plan values
- Deleting all comments for a plan value

Figure 4.65 shows the elements of the example story in this section.

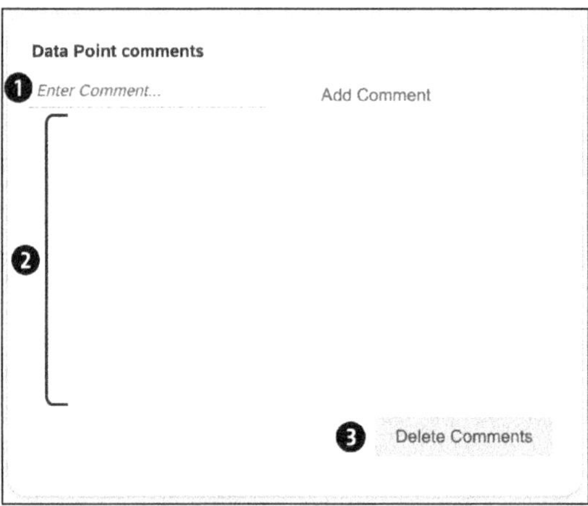

Figure 4.65 Three Areas of This Part of Story

Structure of this part

The structure of this part of the story is divided into three areas:

❶ An input field for entering new comments and a button.

❷ A text field (in Figure 4.65, an empty space below ❶ that is not filled in) that offers the option of displaying data point comments for a table cell.

❸ A widget that is a button intended to allow you to delete the data point comments of a data point.

Show and Add Data Point Comments

In this section, we'll first show you how to read data point comments using a few methods and write them to a text field, and we'll then show you how to add new data point comments.

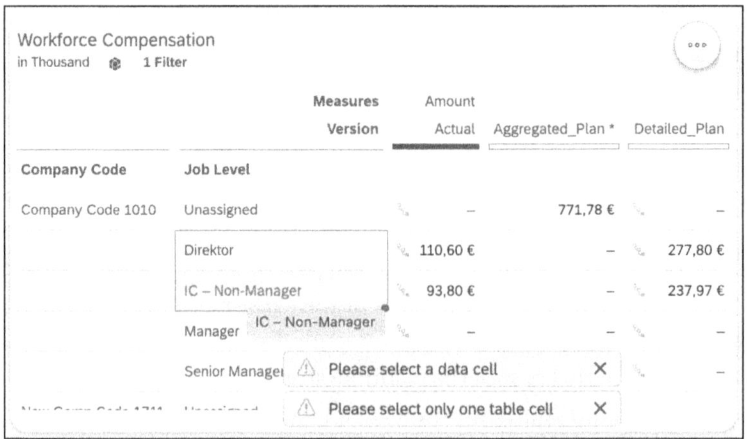

Figure 4.66 Warnings for Incorrect Selection of Table Cell

View comments

As the triggering event for filling the text field with existing comments of a data point, we use the event that is executed when a table cell of the table is selected. We've already discussed this function in Section 4.4.2, and one of the most important aspects of this function is that the user's selection is checked, as data point comments can only be read from individual data cells. Figure 4.66 shows you information that is displayed to users if they make an invalid selection.

How to use the updateComments script function

Another point that we didn't discuss in Section 4.4.2 is the call of the updateComments script function. This is defined via the same script object as the other script functions, but in contrast to the checkSelection script function, it doesn't use an array of the Selection type as an argument, but only an element of the Selection type. A data cell is expected that has previously been validated by other methods, and the return type is void, which means that no value is returned. If data point comments exist for the selected table cell, they are displayed directly above the corresponding text field (see Listing 4.18).

```
var comments = Table_1.getDataSource().getComments().
  getAllComments(selection);
var text = " ";

for (var i = 0; i < comments.length; i++) {
     text = text + "["+ comments[i].createdAt + "] " +
       comments[i].createdBy.displayName + ":\n" +
       comments[i].text + "\n\n";
}

Text_Comments.applyText(text);
```

Listing 4.18 updateComments Method as Part of ScriptObject_1

You can see in Figure 4.67 that the date and time when the comment was created and the display name of the comment creator are displayed in addition to the actual data point comment text. This allows users to understand the context of the comment.

Add data point comment

To add a new comment, you can enter it via the **Enter Comment ...** input field, and the comment will be saved using the addComment method. The triggering event of this method is the onClick event of the **Add Comment** button.

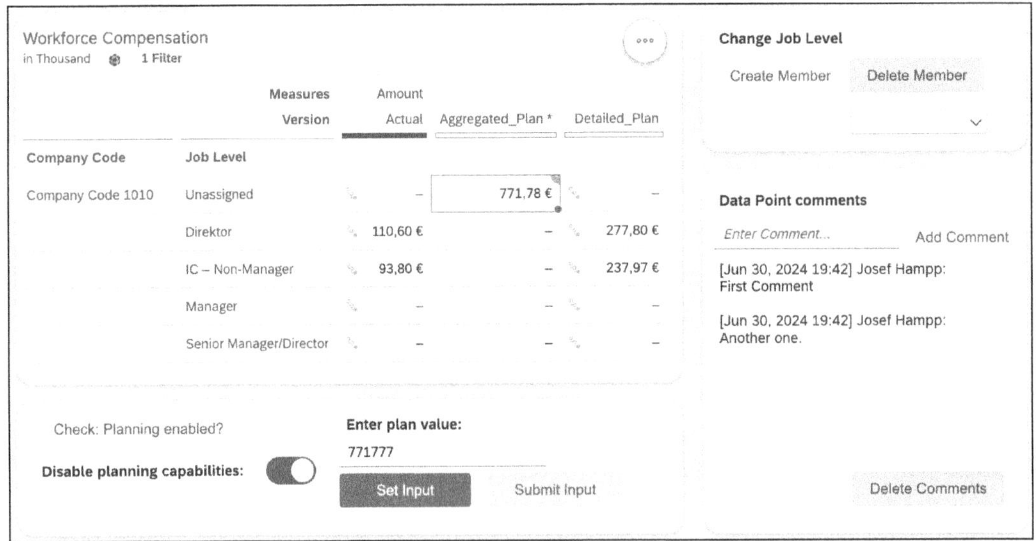

Figure 4.67 Story with Selected Table Cell and Corresponding Data Point Comments

Before the data point comment is added, the script checks whether the selection of the table cell is correct. The `checkSelection` script function is used again for this, and if there are no errors, the data point comment can be added to the data model. Finally, the input field is reset and the `updateComments` script function is called to update the comment field (see Listing 4.19).

```
var comment = InputField_NewComment.getValue();
var selections = Table_1.getSelections();
var resultCheck = ScriptObject_1.checkSelection(selections);

if (!resultCheck) {
     return;
}

Table_1.getDataSource().getComments().addComment(selection, comment);

InputField_NewComment.setValue("");
ScriptObject_1.updateComments(selection);
```

Listing 4.19 onClick Event of Button_CreateComment Button

Deleting Data Point Comments

How to delete a data point comment

In this section, we'll show you how to delete data point comments. To do this, you must select a corresponding table cell, and you can then start the deletion by clicking the **Delete Comments** button. The script is started via the onClick event.

The script is structured similarly to the script in the previous section (see Listing 4.20). First, the user's selection is checked using the checkSelection script function. If the check returns true, the removeAllComments method is executed and removes all comments of a data point that is specified as an argument, and the updateComments script function is then called to update the comment field.

```
var selections = Table_1.getSelections();
var resultCheck = ScriptObject_1.checkSelection(selections);

if (!resultCheck) {
     return;
}

Table_1.getDataSource().getComments().removeAllComments(selection);
ScriptObject_1.updateComments(selection);
```

Listing 4.20 onClick Event of Button_DeleteComments Button

Figure 4.68 shows you the example from Figure 4.67 after the comments of the data point have been removed.

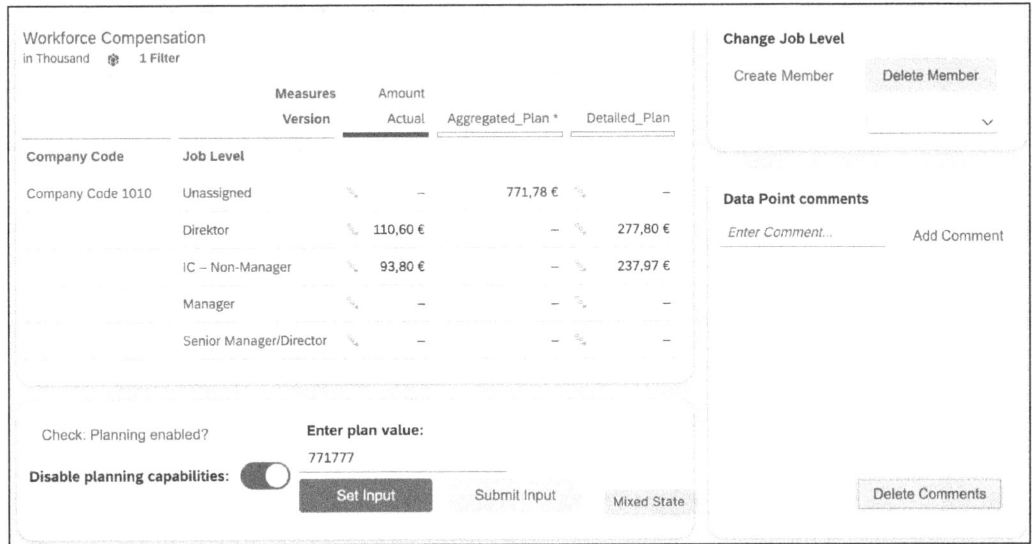

Figure 4.68 Story after Deleting Data Point Comments

4.5 Best Practices

As in the previous chapter, we would like to discuss best practices for planning stories. Of course, you can use all the tips and tricks you've learned for creating BI stories for planning stories, and they are still valid.

Especially in the area of planning, we also recommend that you take a look at the available examples that you can download as business content. These contain common best practices and good presentation methods that you can adopt. In the following sections, we explain best practices using the business content package SAP Operational Workforce Planning.

4.5.1 Entry into the Planning Story

A planning process consists of various phases and steps. For reasons of simplified maintenance and provision, we recommend separating individual steps from one another, and in this context, separation is at the object level.

Depending on the scope, you can create individual stories for each planning phase or planning step. However, to provide users with an easy introduction to the planning process, you should create best practices as an entry point for users.

Figure 4.69 shows you what this entry point could look like. The overview or start page should be structured and not overloaded, and in this example, the most important key figures from Human Resources are displayed in the form of KPI tiles at the top left.

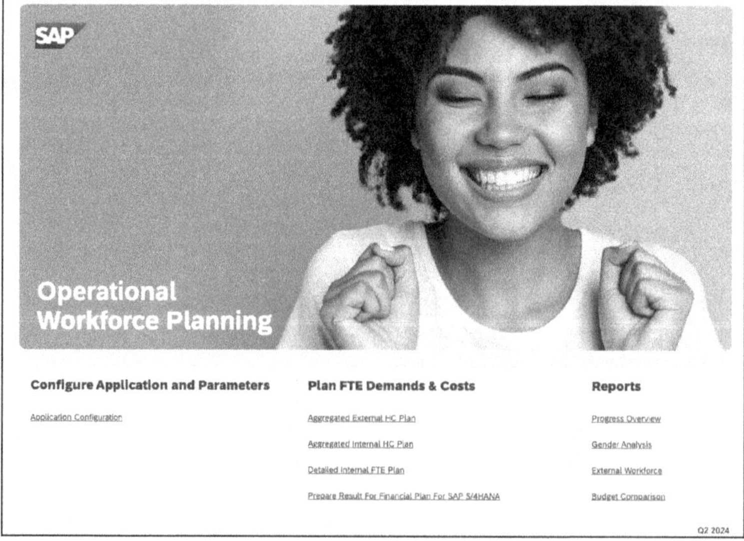

Figure 4.69 Entry Page of Business Content Package for Operational Workforce Planning

The lower section of the application is divided into three visually separate areas:

- **Configure Application and Parameters**
- **Plan FTE Demands & Costs**
- **Reports**

In each area, users of the planning story have the option of jumping to one or more subsequent stories. In this case, users can make the jumps via simple text fields. Instead of the standard functionality for storing hyperlinks in widgets, scripting is used here. A script function is called at the onClick event of the text field in use, and you can see an example of this in Listing 4.21.

```
// Extract a story id and page index at which the story should be
opened from the combination provided.
var storyId = arg_StoryIdPageIndex.substring(0,arg_
StoryIdPageIndex.indexOf(delimiter_storyIdPageIndex));
var pageIndex = arg_StoryIdPageIndex.substring(arg_
StoryIdPageIndex.indexOf(delimiter_storyIdPageIndex)+1);

// Create and fill an array of URL parameters needed to open the
desired story at the selected page.
var urlParameters = ArrayUtils.create(Type.UrlParameter);
urlParameters.push(UrlParameter.create("mode", arg_displayMode));
urlParameters.push(UrlParameter.create("page", pageIndex));


//Predictive Toggle URL Parameter is disabled but can be added here
as needed.
/*
if (toggleIsPredictiveEnabled){
    urlParameters.push(UrlParameter.create('p_
toggleIsPredictiveEnabled', 'true'));
}
*/

NavigationUtils.openStory(storyId, '', urlParameters, false);
```

Listing 4.21 navigateToStory Script Function

The script creates an empty object of the UrlParameter type, and the ID of the story is then taken from a script object that is the target of the jump. The script element is defined with the story's ID when the entry story is started, and URL parameters are then defined and added to the empty object. Finally, the target story is opened using the openStory method.

The different stories are stored in a folder in the SAP Analytics Cloud file system (see Figure 4.70).

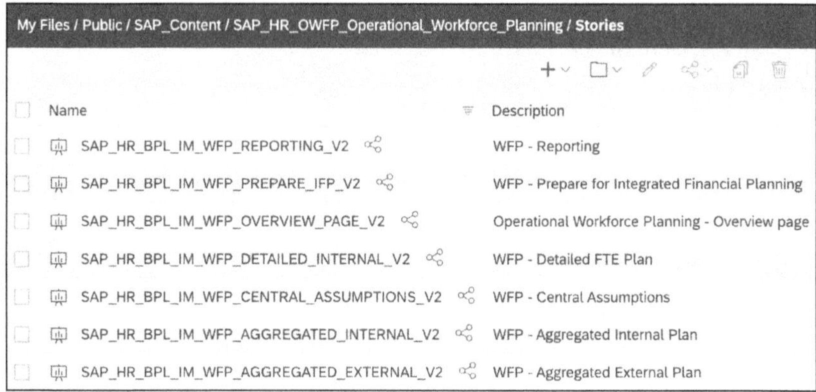

Figure 4.70 Separate Stories of Business Content Package

This is irrelevant for users but easier to manage for the administrators of the planning story.

4.5.2 Assistance for Users

To support your users in the planning process, you can implement help functions, some of which are shown in Figure 4.71.

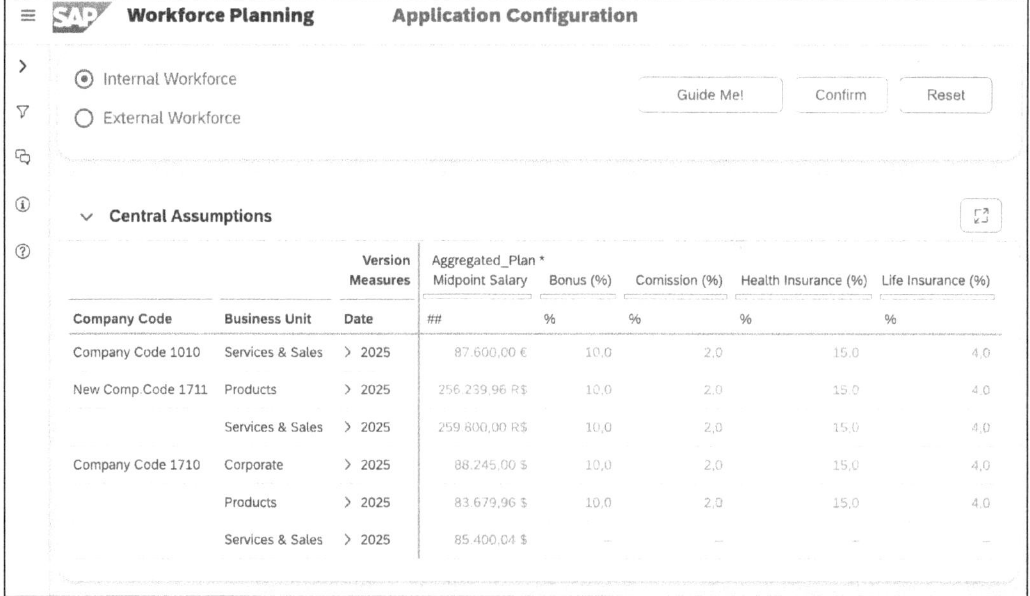

Figure 4.71 Assistance for Users

First, we'll look at the sidebar, which is used in every single story of the business content except the entry page. Figure 4.72 shows the sidebar in detail. It offers several features, such as filtering and navigation options. You can also jump to external pages from it, for example, to ask a question or read information on the planning process.

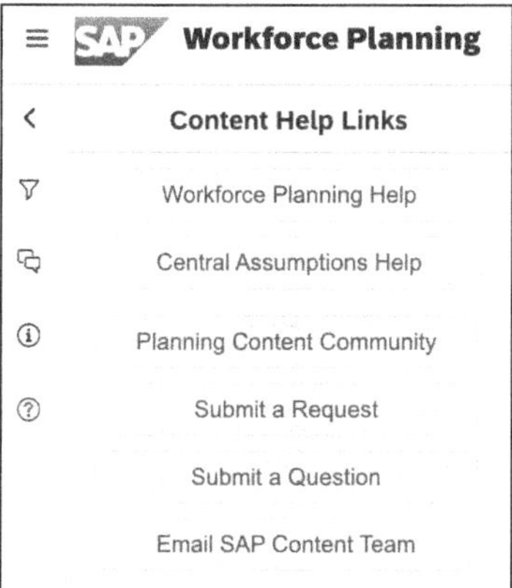

Figure 4.72 Sidebar of Business Content

Planning is also part of the sidebar of the story, and it usually consists of several steps. In the **Instructions** area, you'll find step-by-step instructions for your users, and these ensure that even occasional users have the information they need to carry out the individual planning steps correctly. Figure 4.73 shows you the **Instructions** area of the business content.

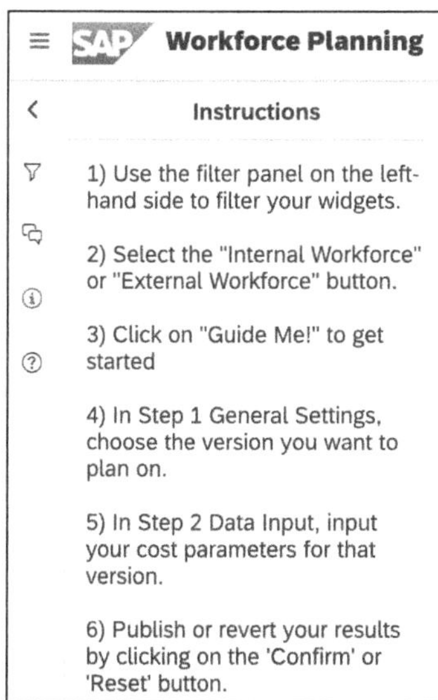

Figure 4.73 Step-by-Step Instructions

Finally, there are buttons for individual actions. Figure 4.74 shows the buttons that are used in the **Application Configuration** story, which are color coded green and red to make it clear to users which functions they have.

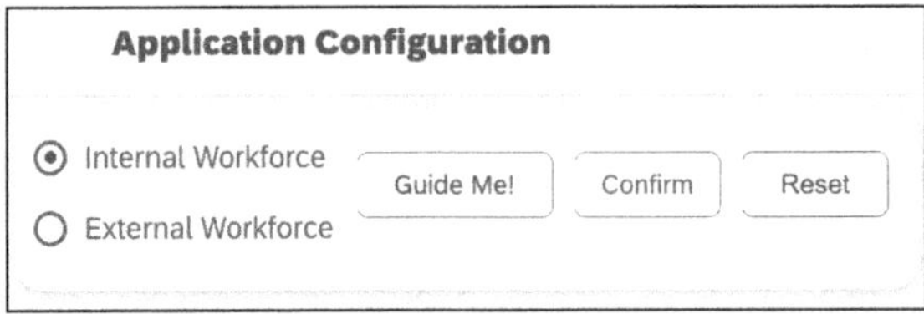

Figure 4.74 Buttons as Interaction Options for Users

In addition to the step-by-step instructions, a guided process was implemented when creating the business content. After you click the **Guide Me!** button, a popup opens and provides you with a focused view of the steps that you can take as the executor of the story. Figure 4.75 shows you this popup.

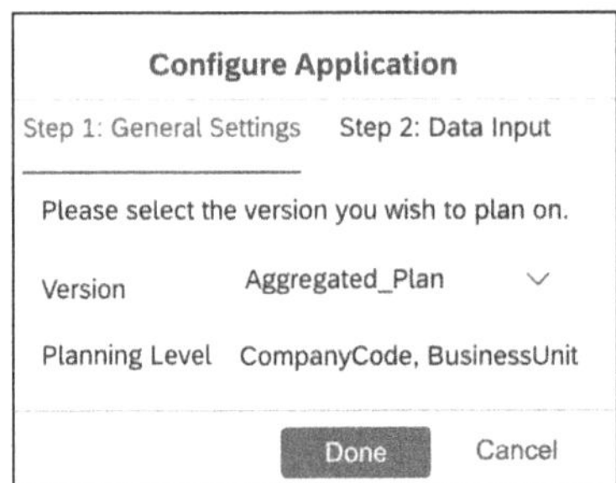

Figure 4.75 Popup When Clicking Guide Me!

4.6 Summary

In this chapter, we gave you an overview of the topic of application development for planning. Having covered specific widgets and APIs for planning, we also showed you a sample story with some of these elements and explained it step by step. We then highlighted some best practices for this topic.

Now that you have an understanding of story design with scripting in the areas of BI and planning, we'll show you in the next chapter how to create your own widgets.

Chapter 5
Custom Widgets

In this chapter, you'll learn how to develop your own widgets in addition to the widgets provided by SAP for stories and integrate them into your stories.

5

What you can expect

SAP Analytics Cloud offers a wide range of standard functions and widgets that you can use directly to create stories. This standard repertoire usually covers the majority of all requirements, but there are special requirements that can't be met with the standard. To enable you to fulfill all of your stakeholders' wishes, SAP Analytics Cloud is also open to new *custom widgets* that you can develop yourself and then simply use in your story.

In this chapter, we'll teach you all about custom widgets and how to use them. In Section 5.1, "What Are Custom Widgets?" we'll give you an overview of what custom widgets are and go into the restrictions to be observed and the hosting. In Section 5.2, "Setup of Custom Widgets," we'll teach you the structure of widgets and the files required for them. You must program HTML and JavaScript code to develop your own widgets, so in Section 5.3, "Creating, Uploading, and Removing a Custom Widget," we'll clarify the requirements and compare the various architecture and deployment options. Then, in Section 5.4, "An Example of a Custom Widget: The Colorful Box," we'll use a step-by-step example to show you how to develop your own widget and integrate it into a story. In Section 5.5, "Custom Widgets on Mobile Devices," we'll describe what you need to bear in mind when using your widgets on mobile devices. We dedicate Section 5.6, "Exporting Custom Widgets," to the export of custom widgets, and in Section 5.7. "Best Practices," we conclude by discussing best practices for custom widgets.

5.1 What Are Custom Widgets?

Custom widgets

With custom widgets, SAP Analytics Cloud offers a framework that allows you to use your own *custom widgets* to extend the selection of widgets provided by the story designer. You'll need custom widgets to create specific visualizations, elements of the user interface, and functions in your story that aren't provided by the standard widgets of SAP Analytics Cloud.

For story designers, custom widgets are part of the holistic user experience in SAP Analytics Cloud and can be fully integrated into the solution. Like all other widgets, custom widgets offer the following functions:

Similarities to standard widgets

- They are listed in the widget menu bar, and you can add them to your application from there.
- They can be moved and resized in the canvas area of your story.
- They are displayed in the widget outline.
- They can provide scripting methods for the story designer scripting language.
- They can be defined via properties that you can configure while editing your story in the designer (in the builder and format area).

Technical background

From a technical point of view, custom widgets are small web applications. In its simplest form, a custom widget only consists of a JSON file and a JavaScript file. More complex custom widgets, on the other hand, consist of a large number of files (image files, CSS files, JavaScript files, etc.).

Due to the simple structure of custom widgets, you don't need any special programs to create them; you can even create simple widgets in a text editor. Nevertheless, the use of a design time is recommended, especially for more complex widgets. This design time should be able to work with JavaScript files.

Implementation as a web component

Custom widgets are implemented as *web components*, which are used to integrate custom HTML elements (so-called *custom elements*) into the HTML Document Object Model (DOM) without affecting the rest of the HTML DOM. This ensures, among other things, that the styling and rendering of custom elements is strictly isolated from the rest of the HTML DOM. This is achieved by a separation in the shadow DOM, where the HTML DOM of a custom widget is separated from the HTML DOM of the website. Web components consist of HTML and JavaScript, and CSS can be used as an add-on.

As in the previous chapter, you don't need any specialist knowledge for this chapter. Knowledge of web components isn't a prerequisite, and we explain all relevant ideas and concepts of web components in the text and demonstrate them using sample code.

5.1.1 Restrictions

Restrictions on use

Custom widgets have limitations that you should be aware of when developing and using them. For example, custom widgets are only supported in Microsoft Edge (version 79 and higher) and Google Chrome. There are also functional restrictions on the development of custom widgets. You can't currently use the following functions:

- Data blending
- Data change insights
- Commenting
- Pause refresh
- Scheduling of publications
- Searching to insight
- Theme and CSS
- Translation
- Viewport loading

[«]

Enabling Linked Analysis and Bookmarking in the Optimized Design Experience

Linked analytics and bookmarks for custom widgets are supported in the optimized design experience. This was previously not possible for stories using Analytics Designer.

In specific cases, you need to perform further actions to enable linked analysis. For more information, refer to "Enabling Linked Analysis for Your Custom Widget (Optimized Story Experience)" (see *http://s-prs.co/v594404*).

To enable bookmark support for your custom widget, you need to perform further actions in its JSON file. For more information, refer to "Configuring Bookmark Support for Your Custom Widget (Optimized Story Experience)" (see *http://s-prs.co/v594405*).

Restriction for data connections

There are also restrictions when using custom widgets with data connections, which are only supported in optimized view mode. There are also restrictions when using APIs, as the following APIs are not supported by custom widgets with data connections:

- `getDataSource().getComments`
- `getDataSource().load`
- `getDataSource().getDataExplorer`
- `getDataSource().getDataSelections`

[«]

Validity of the Restrictions

All restrictions described in this section were valid when this book was written, but changes may have been made in the meantime. You can view the currently valid restrictions in the SAP Help Portal at *http://s-prs.co/v594406*.

5.1.2 Hosting Custom Widgets

Providing custom widgets

To host custom widgets, you need two types of files: a JSON file and so-called resource files. All descriptive data (the metadata on the custom widget) is contained in the JSON file, all components of the custom widget are defined there, and the resource files are referenced there via URLs.

Resource files

Resource files are all other files that are required for custom widgets from a functional perspective. These include JavaScript files, CSS files, and HTML files.

The JSON file of the custom widget must be uploaded directly to the SAP Analytics Cloud Tenant.

You have two options for where to upload and host the resource files:

- Your own HTTP web server
- SAP Analytics Cloud
- We'll discuss these options in the following sections.

Hosting via your own HTTP Web server

HTTP Web Server

You can upload the resource files of the custom widget to an HTTP server, which must have HTTPS enabled. The resource files are not executed or processed on the web server, and they are also static. The web server only serves as a simple storage location for the files, and the JSON file of the custom widget refers to these resource files via URL references.

Figure 5.1 shows you an example of how to host a custom widget via a public web server.

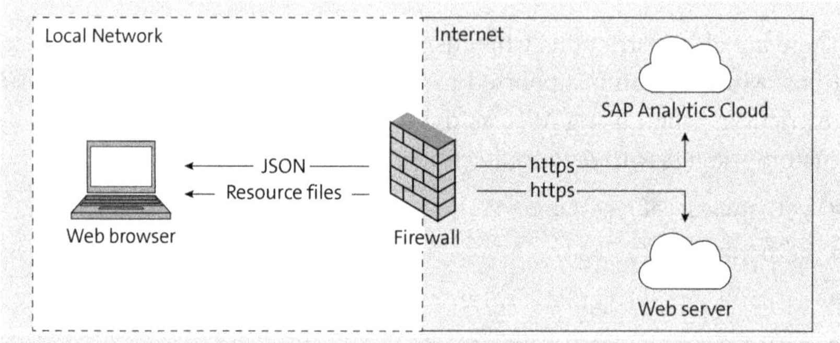

Figure 5.1 Hosting of Custom Widgets

How the custom widget framework works

If a custom widget is rendered in the web browser, the URL of the resource file is transferred to the browser. This is done via the *custom widget framework*, which takes the URL from the JSON. The browser uses this information to request the resource files from the web server, and since the resource files are simple file types, they should be provided in every web browser. The custom widget framework doesn't place any special requirements on the web server and doesn't offer any authentication or authorization functions or support for session cookies or similar.

The resource files on the web server are requested by the custom widget both when the story is created in the design time and during runtime, when the story is executed. If the resource files are only used for the builder or styling panel, these files are only requested by the web server during the design time.

Hosting variants

The web server can be either public or private. In this context, *private* means that it's accessible within the company or the company network. The only requirement for a private web server is that the resource files can be provided by the user's browser under the URL specified in the JSON file. For example, you can operate a *Node.js* server with the HTTPS server module or an Apache web server on your local system as a private web server. The advantage of a private web server is that it is only accessible within your company and is therefore particularly suitable for developing your own custom widgets.

[«]

Recapitulation of the Mode of Operation

The SAP Analytics Cloud tenant doesn't connect directly to the web server to request resource files; instead, the web browser is always used for this purpose.

If you save a story on your SAP Analytics Cloud tenant, only a reference to the custom widget (specifically, the JSON file of the story) is saved with the story. Resource files are never stored in stories.

SAP Analytics Cloud

To host a custom widget on SAP Analytics Cloud, you have to adjust the URL file paths of the resource files in the JSON file. These should start with a forward slash. Later on in this chapter, Listing 5.12 will demonstrate hosting on a web server. The URL is written out in its entirety, and using the example of *coloredbox_main.js*, the URL property is adjusted as shown in Listing 5.1.

```
{
"kind": "main",
"tag": "com-sap-sample-coloredbox",
"url": "/coloredbox_main.js",
"integrity": "",
"ignoreIntegrity": true
},
```

Listing 5.1 Adjusted URL for Hosting Resource Files on SAP Analytics Cloud

To upload the resource files to the SAP Analytics Cloud tenant, you must compress the files in a zip file.

After you upload the JSON file of the custom widget, SAP Analytics Cloud recognizes from the definition of the `URL` property that hosting via the tenant is requested, and it instructs you to provide the zip file. The exact process is described in Section 5.3.3.

If you decide to go for this option, there are a few things you should bear in mind:

- The file extension of the compressed file should be *.zip*.
- The zip file should not be larger than 5 MB.
- The zip file only supports the following resource files: the JavaScript web component, the JavaScript web component of the styling panel, the JavaScript web component of the builder panel, and icon. CSS and HTML files are not supported.
- Icon only supports PNG and JPG files.
- Subfolders in the zip file are not supported.
- A maximum of twenty-five custom widgets with SAP-hosted resources are allowed per client.

You also need the **Create** permission enabled for **Custom Widget** to create a new custom widget on SAP Analytics Cloud.

[!]

SAP Analytics Cloud Based on SAP BTP, Neo Environment

Hosting resource files is only supported for SAP Analytics Cloud tenants running on SAP BTP, Cloud Foundry environment.

5.2 Setup of Custom Widgets

Basic structure

Custom widgets are made up of the following files:

- The JSON custom widget
- The JavaScript web component
- The JavaScript web component of the styling panel
- The JavaScript web component of the builder panel
- Icon
- Other files

Explanation of the files

The *JSON custom widget* document is required, and it specifies the custom widget.

The *JavaScript web component* implements the custom element of the custom widget. This file is also required.

The *JavaScript web component of the styling panel* is optional. It implements the custom element of the styling panel of the custom widget.

The same applies to the *JavaScript web component of the builder panel*, which implements the custom element of the builder panel of the custom widget. The *icon* is also optional. It is used to represent the custom widget and has a format of 16 × 16 pixels.

Other files (JavaScript, CSS, and images) can also be used, depending on the custom widget implemented. As a result, such files are also optional and not essential components of every custom widget.

These files are explained in more detail in the following sections.

5.2.1 JSON Custom Widget Reference Document

Reference document

The *JSON reference document* (root object) specifies the custom widget and refers to its components. All valid properties of the JSON reference document are explained in the following sections. If there are other properties present in this reference document, the custom widget will render `invalid`.

Master element

The *root object* of the JSON custom widget reference document specifies the custom widget. In Table 5.1, you can see the properties that you can use.

Property	Required/ Optional	Type	Description
`id`	Required	`String`	The unique ID of the custom widget.
`name`	Required	`String`	The name of the custom widget.
`properties`	Required	`Properties` object	The properties of the custom widget.
`version`	Required	`String`	The version of the custom widget.
`webcomponents`	Required	Array of `Webcomponents` objects	The web components of the custom widget.
`dataBindings`	Optional	`DataBindings` object	The data bindings of the custom widget. You can define your own data binding with an array of `feed` objects.

Table 5.1 JSON Properties of Root Object

Property	Required/ Optional	Type	Description
description	Optional	String	The description of the custom widget.
eula	Optional	String	The end user license agreement.
events	Optional	Events object	The events of the custom widget.
icon	Optional	String	The URL of the icon of the custom widget.
imports	Optional	Array of String objects	The type libraries used by the custom widget.
license	Optional	String	The license.
methods	Optional	Methods object	The methods of the custom widget.
newInstancePrefix	Optional	String	The prefix of a new custom widget instance.
supportsBookmark (optimized design experience)	Optional	Boolean	This indicates whether the custom widget supports bookmarking.
supportsExport	Optional	Boolean	This indicates whether the custom widget supports PDF exporting.
supportsLinkedAnalysisFilterOnSelection (optimized design experience)	Optional	Boolean	This indicates whether the custom widget supports linked analysis based on a filter on data point selection. In addition, within your JavaScript web component file, you need to use data binding's linked analysis APIs so that a selected data point from the custom widget can filter other widgets.
supportsMobile	Optional	Boolean	This indicates whether the custom widget is supported on mobile devices.

Table 5.1 JSON Properties of Root Object (Cont.)

Property	Required/ Optional	Type	Description
types	Optional	Types object	Custom types of the custom widget. You can define your own data structures and enumerations.
vendor	Optional	String	Provider.

Table 5.1 JSON Properties of Root Object (Cont.)

[+]

Icon

You can also specify a data URL as the source for the icon. With a data URL, you can link icons directly in the JSON file of the custom widget that would otherwise be requested via the network.

In the following sections, we'll explain the different JSON objects and their properties.

Webcomponents Object

A `Webcomponents` object specifies a web component of a custom widget. It's defined by the JSON properties in Table 5.2.

Properties of the Webcomponents object

Property	Required/ Optional	Type	Description
integrity	Required	String	The hash value of the JavaScript file of the web component. The hash value has the format *hashAlgorithm-hashValue*.
kind	Required	String	This describes what type of web component this object represents. Possible values are as follows: ■ builder (the builder panel of the custom widget) ■ main (the actual custom widget) ■ styling (the styling panel of the custom widget)
tag	Required	String	The unique name of the custom element (the HTML tag) of this web component.
url	Required	String	The URL of the JavaScript file of the web component.

Table 5.2 JSON Properties of Webcomponents Object

Property	Required/ Optional	Type	Description
ignoreIn-tegrity	Optional	String	This specifies whether the integrity check can be ignored.
type	Optional	String	This can be used to modularize JavaScript.

Table 5.2 JSON Properties of Webcomponents Object (Cont.)

Tag

To distinguish a custom widget from a regular HTML element, the unique name of the custom widget must contain at least one hyphen (-).

[+]

Creating Tags

You can derive the tag from the ID of the custom widget. To do this, replace all dots with hyphens. This creates a unique tag.

By adding the major version number to the tag, you can avoid confusion among multiple versions of a custom widget that are added to the story designer at the same time.

Properties Object

Properties of the Properties object

A `Properties` object specifies the properties of a custom widget. Each JSON property name of a `Properties` object is the name of a property of the custom widget, and its value is an object with various JSON properties (see Table 5.3).

Properties	Required/ Optional	Type	Description
type	Required	String	The type of property of the custom widget. Possible values are as follows: ▪ `Boolean` ▪ `Integer` ▪ `Number` ▪ `String` ▪ `Boolean[]` ▪ `Integer[]` ▪ `Number[]` ▪ `String[]` ▪ Certain script API data types ▪ Simple object types

Table 5.3 JSON Properties of Properties Object

Properties	Required/ Optional	Type	Description
`default`	Optional	Depends on type property	The default value of the property of the custom widget. One of the following JavaScript types is a possible value: ▪ `Boolean` ▪ `Number` ▪ `String` ▪ Array of `Boolean` ▪ Array of `Number` ▪ Array of `String`
`description`	Optional	`String`	The description of the property of the custom widget.

Table 5.3 JSON Properties of Properties Object (Cont.)

[«]

Changing Properties by Using Script Methods

If you want to modify a property that is an array by using a script method, you must always reassign the modified array to the property after the modification.

Methods Object

Properties of the Methods object

A `Methods` object specifies the script methods of a custom widget. Each JSON property name of a `Methods` object is the name of a method of a custom widget script, and it's defined by the JSON properties in Table 5.4.

Properties	Required/ Optional	Type	Description
`description`	Required	`String`	The description of the script method.
`body`	Optional	`String`	The implementation of the script method.
`parameters`	Optional	Array of parameter objects	The parameter of the script method. If the script method has no parameters, omit this property.
`returnType`	Optional	`String`	The return type of the script method. If the script method has no return value, omit this property.

Table 5.4 JSON Properties of Methods Object

[»]

Body

If you omit the body property, the custom widget framework searches for a so-called native JavaScript function with this name in the JavaScript file of the web component whose kind property has the main value and calls it.

When you change a property of a custom widget in a native JavaScript function, you need to send a custom propertiesChanged event to notify the custom widget framework of the change. If you don't do this, it can lead to outdated property values if you call it in a script method. For more information and a code example, see Section 5.4.3.

Parameter Object

Properties of the Parameters object

A Parameter object specifies one parameter of a script method of a custom widget. It's defined by the JSON properties in Table 5.5.

Properties	Required/ Optional	Type	Description
name	Required	String	The name of the parameter.
type	Required	String	The type of property of the custom widget. Possible values are as follows: ■ Boolean ■ Integer ■ Number ■ String ■ Boolean[] ■ Integer[] ■ Number[] ■ String[] ■ Certain script API data types ■ Simple object types
description	Optional	String	The description of the parameter.

Table 5.5 JSON Properties of Parameter Object

Events Object

Properties of the Events object

An Events object specifies the events of a custom widget. Each JSON property name of an Events object is the name of an event of a custom widget, and the Events object has the JSON property in Table 5.6.

Properties	Required/Optional	Type	Description
description	Optional	String	The description of the event.

Table 5.6 JSON Property of Event Object

Types Object

A Types object specifies the custom types of a custom widget. You can use custom data structures or custom enumerations as custom types, and within the JSON custom widget, you can use custom types as types for properties, arguments, and return types. In stories, you can use custom types as types for global variables, return types, and script object function arguments.

Unambiguous naming

Each JSON property name of a Types object is the name of a custom type. To avoid multiple assignments of custom type names, each name of a custom type is made unique internally by using a qualified name that contains the ID of the custom widget. You don't need to specify the qualified name.

[«]

Predefined Type Names

If names for custom types are used in Methods, Parameters, and Properties objects, they take precedence over the predefined type names.

Properties of a custom data structure

You can use the JSON properties in Table 5.7 to specify custom data structures.

Properties	Required/ Optional	Type	Description
description	Required	String	The description of the data structure.
properties	Required	Array of CustomTypeProperty objects	The properties of the data structure.
body	Optional	String	The implementation of the script method.
extends	Optional	String	The data structure to be extended. You can extend your own custom data structures, but you can't extend custom enumerations and simple types (Boolean, Integer, Number, and String).

Table 5.7 JSON Properties of Custom Data Structure

Properties of a custom enumeration type

You can use the JSON properties in Table 5.8 to specify custom enumeration types.

Properties	Required/ Optional	Type	Description
description	Required	String	The description of the custom enumeration type.
properties	Required	Array of CustomTypeProperty objects	The properties of the custom enumeration type.
extends	Optional	String	The data structure to be extended. You can extend your own custom data structures, but you can't extend custom enumerations and simple types (Boolean, Integer, Number, and String).

Table 5.8 JSON Properties of Custom Enumeration Type

CustomTypeProperty Object

Properties of the CustomType Property object

A CustomTypeProperty object specifies one property of a custom type of a custom widget, and it has the JSON properties listed in Table 5.9.

Properties	Required/ Optional	Type	Description
name	Required	String	The name of the CustomTypeProperty object.
type	Required	String	The type of the property. You can use a simple type (Boolean, Integer, Number, or String) an enumeration, or a type that can be used as an argument type or as an object expression.
description	Optional	String	The description of the CustomTypeProperty object.
mandatory	Optional	Boolean	This specifies whether this property is mandatory. If you want to create instances of this type, you must specify this property.

Table 5.9 JSON Properties of CustomTypeProperty Object

[«]

Types

In the API reference, you'll find all types that you can use as object expressions or argument types. To do this, search for "API Reference Guide" in the SAP Help Portal. If types can be used as argument types, they are marked with a round icon and a *T*.

If types can be used as object expressions, the text "Can be passed as a JSON object to method arguments" appears at the beginning of their class documentation.

CustomEnumValue Object

Properties of the CustomEnumValue object

A `CustomEnumValue` object specifies one value of a custom enumeration type of a custom widget, and it has the JSON properties listed in Table 5.10.

Properties	Required/Optional	Type	Description
`name`	Required	`String`	The name of the `CustomEnumValue` object.
`description`	Optional	`String`	The description of the `CustomEnumValue` object.

Table 5.10 JSON Properties of CustomEnumValue Object

DataBindings Object

Properties of the dataBindings object

A `dataBindings` object specifies the data binding of a custom widget in the `dataBindings` property. Only one object can be defined, and if several objects are specified, only the first object is used. The `dataBindings` object has the JSON properties in Table 5.11.

Properties	Required/Optional	Type	Description
`feeds`	Required	Array of `feed` objects	The feed of the `dataBindings` object.

Table 5.11 JSON Properties of dataBindings Object

[«]

Names of the dataBindings Objects

The name of an object is represented by the key of the `dataBindings` property.

If you define a dataBindings object in the JSON file, a default builder panel is displayed in the design time. If you have defined your own builder panel via the Webcomponent object, the default builder panel doesn't appear.

Feed Object

Properties of the feed object

The feed object has the JSON properties listed in Table 5.12.

Properties	Required/Optional	Type	Description
id	Required	String	The unique ID of the feed object.
type	Required	String	This defines what you can add to the corresponding feed object. Possible values are dimension or mainStructureMember (also known as measure).
description	Optional	String	The description of the feed object.

Table 5.12 JSON Properties of Feed Object

Simple Object Types

In addition to simple types (Boolean, Integer, Number, and String), script API data types, custom types, and arrays of these types, you can use a *simple object type* of the Object<type> form. This type represents a JavaScript object with keys of the String type and values of the type<type> form. For example, an Object<String> type represents a JavaScript object with keys of the String type and values of the String type.

The following elements of the Type<type> form are supported:

- Simple types (Boolean, Integer, Number, and String)
- Custom types

Example of simple object types

The following example shows you how to use an object of the Object<Number> type. This object has keys of the String type and values of the Number type.

The object of the Object<Number> type is first defined in the JSON file of the custom widget (see Listing 5.2).

```
{
    ...
    "properties": {
        "myNumbers": {
            "type": "Object<number>",
            "description": "A collection of number object"
```

```
        }
    },
    "methods": {
        "putNumber": {
            "parameters": [
                {"name": "name", "type": "string"},
                {"name": "value", "type": "number"}
            ],
            "description": "Adds a number, using a name, to the
collection."
        },
        "getNumber": {
            "parameters": [
                {"name": "name", "type": "string"}
            ],
            "returnType": "number",
            "description": "Returns a number, using a name, from the
collection."
        }
    }
}
```

Listing 5.2 JSON Custom Widget

In Listing 5.3, methods are defined in the JavaScript web component.

```
putNumber (name, value) {
  this.myNumbers = this.myNumbers || {};
  this.myNumbers[name] = value;
}

getNumber (name) {
  return this.myNumbers[name];
}
```

Listing 5.3 JavaScript Web Component

The code example in Listing 5.4 creates two objects of the `Object<number>` type in the first step using the `put` method. Subsequently, one of these objects is output again using the `get` method.

```
var number1 = 1;
var number2 = 2;

customWidget.put("n1", number1);
```

```
customWidget.put("n2", number2);
```

```
var n = customWidget.get("n2");
```

Listing 5.4 Sample Script

5.2.2 JavaScript Web Component

As explained in the previous sections, a custom widget consists of one or more web components. Each web component is implemented in a JavaScript web component file, which defines a custom widget and implements its JavaScript API. You can see an example of this in Section 5.5. In the current section, we'll explain more about web components, their lifecycle methods, script APIs, and the use of data bindings.

Lifecycle of a Web Component

The custom widget framework calls certain JavaScript functions of the web component in a predefined order during the usage time of a custom widget. If you want to control the behavior of the custom widget, you can implement these JavaScript functions yourself. You can find a detailed explanation of each of these functions in the next section on the JavaScript API web component.

Function calls when rendering for the first time

When the custom widget is rendered for the first time, the custom widget framework calls the following sequence of JavaScript functions:

1. `constructor()`
2. `onCustomWidgetBeforeUpdate()`
3. Property setter functions for updating the properties of the custom widget (if available)
4. `onCustomWidgetAfterUpdate()`
5. `connectedCallback()`

Function calls during the update

If the custom widget is updated, the custom widget framework calls the following sequence of JavaScript functions:

1. `onCustomWidgetBeforeUpdate()`
2. Property setter functions for updating the properties of the custom widget (if available)
3. `onCustomWidgetAfterUpdate()`

Function calls during deletion

If the custom widget is removed from the canvas or the story is closed, the custom widget framework calls the following sequence of JavaScript functions:

1. `onCustomWidgetDestroy()`
2. `disconnectedCallback()`

> **Execution of the onCustomWidgetDestroy Function**
>
> The `onCustomWidgetDestroy` function isn't executed under two conditions:
>
> - If the visibility of the custom widget is set to `false`
> - If the custom widget is an invisible part of a container

Function calls during processing

If the size of the custom widget on the canvas is changed, the custom widget framework calls the `onCustomWidgetResize()` JavaScript function.

> **Moving the Custom Widget**
>
> If you want to drag a custom widget onto the canvas during design time, the custom widget will be cloned. This provides you with an object to move, and it calls the constructor of the custom widget as well as the `connectedCallback` and `disconnectedCallback` callbacks.

JavaScript API Web Component

The following functions are available to you via the JavaScript API of the web component of a custom widget.

The constructor() function

You can implement the `constructor()` function to execute JavaScript code when the web component is initialized. This function is the counterpart to the `onCustomComponentDestroy()` function.

The onCustomWidgetBeforeUpdate (oChangedProperties) function

You can implement the `onCustomWidgetBeforeUpdate(oChangedProperties)` function to execute JavaScript code before properties of the custom widget are updated. The `oChangedProperties` argument is a JavaScript object and contains the changed properties as key-value pairs. The name of the property is defined by the key, and the value of the pair is the changed value of the property. If the custom widget framework calls the function for the first time, a complete list of the properties is passed in `oChangedProperties`. If the property value of a property is defined in the JSON file of the custom widget, it's the default value; otherwise, it's of the `undefined` type.

> **Additional Properties**
>
> In a special case, additional values can be passed in `oChangedProperties`. When you're calling web components that implement the actual custom widget (with `kind` = `main` in the JSON file of the custom widget), the values in Table 5.13 are also passed on the first call.

Property	Type	Description
designMode	Boolean	This indicates whether the custom widget is currently running in the design mode.
Event name (for example, onClick)	Boolean	This indicates whether an event script is assigned to this event of the custom widget.
widgetName	String	The name of the widget.

Table 5.13 Additional JSON Properties of oChangedProperties

The onCustomWidgetAfterUpdate (oChangedProperties)

You can implement the `onCustomWidgetBeforeUpdate(oChangedProperties)` function to execute JavaScript code after the properties of the custom widget have been updated.

The onChangedProperties argument is a JavaScript object and contains the changed properties as key-value pairs. The name of the property is defined by the key, and the value of the pair is the changed value of the property.

When the custom widget framework calls the function for the first time, a complete list of properties is passed in `oChangedProperties`. If the property value of a property is defined in the JSON file of the custom widget, it's the default value; otherwise, it's of the `undefined` type.

The connectedCallback() function

You can implement the `connectedCallback()` function to execute JavaScript code when your web component of the custom widget is connected to the HTML DOM of the web page. The following actions can trigger the function:

- Moving the custom widget in the design time, thus cloning the custom widget to provide an object for dragging
- Moving the custom widget from one container to another in the design time
- Rendering the custom widget for the first time in the design time and at runtime
- Moving the custom widget from one container to another at runtime, using the `Widget.moveWidget()` script method

Best Practices for the connectedCallback Function

For custom widgets with a lot of status data, we recommend that you implement the `connectedCallback` function so that it can be called at any time. This allows you to restore the state of the custom widget. For simpler custom widgets with little status data, it may also be easier to re-create the custom widget, depending on the situation; you can use the disconnectedCallback function for this, and we'll discuss it in the next paragraph.

The disconnectedCallback() function

You can implement the disconnectedCallback() function to execute JavaScript code when your web component of the custom widget is disconnected from the HTML DOM of the web page. The following actions can trigger the function:

- Removing the custom widget from the canvas in the design time
- Dragging the custom widget into the canvas in the design time, which clones the custom widget to provide an object for dragging
- Moving the custom widget from one container to another in the design time
- Closing the story in the design time and during runtime
- Moving the custom widget from one container to another at runtime by using the Widget.moveWidget() script method

[+]

Best Practices for the disconnectedCallback Function

For custom widgets with a lot of status data, we recommend that you implement the disconnectedCallback function so that it can be called at any time. This allows you to save the state of the custom widget. The permanent removal of a custom widget is an exception here. For simpler custom widgets with little status data, it may not be necessary to save the status data, depending on the situation.

The onCustomWidgetDestroy() function

You can use the onCustomWidgetDestroy() function to execute JavaScript code when the custom widget is destroyed. This function is the counterpart of the constructor() function.

The onCustomWidgetResize() function

You can implement the onCustomWidgetResize(width,height) function to execute JavaScript code when the size of the custom widget is changed. The width argument specifies the width of the custom widget in pixels, and the height argument specifies the height.

[«]

Additional Information on the Function

The onCustomWidgetResize function is only called for web components with a value of kind = main in the JSON file of the custom widget. For performance reasons with regard to resizing, this function is called at certain time intervals, and the length of the time intervals can change.

When You Should Not Use This Function

If the custom widget isn't affected by the resizing, you should not implement this function. This will prevent the (possibly empty) function from being called repeatedly during resizing.

The property setter function

To define the properties of a custom widget, you can use *property setter functions*, which consist of the keyword `set` followed by the name of the property. The new property value is specified as the argument of the function. The following example shows the function for the `color` property:

```
set color(newColor) {
    this.style["background-color"] = newColor;
}
```

Property setter functions are optional. They are called by the custom widget framework if they are implemented, and if they are not implemented, they are skipped.

[»]

Getter Functions

Getter functions that are implemented in a similar syntax with the keyword `get` are not recognized by the custom widget framework and are ignored.

Alternative to Property Setter Functions

Depending on the functional design of a custom widget, there is also the approach of dispensing with property setter functions and instead implementing the updating of property values in the `onCustomWidgetAfterUpdate` function.

Script API Data Types in JavaScript Functions

When implementing a web component of a custom widget, you can use certain *script API data types* in native JavaScript functions. You can then use these as function arguments or return values, for example, and you can also call functions that are provided by these script API data types.

A native JavaScript function is a JavaScript function in a web component of your custom widget, and there are also script methods that your custom widget provides in the story script editor.

Specifications for the use of script API data types

The following rules apply when using script API data types in native JavaScript functions:

- You may only call documented functions that are provided by a script API data type.
- You must prepare your code so that it correctly accepts the return value of a call for a script API data type and returns it asynchronously.

To guarantee the correct retrieval of the function result, use one of the following programming patterns in your native JavaScript function:

1. Use the keywords `async` and `await`. In the example in Listing 5.5, you can see how to call the `getText` function of a button that was passed as a function argument to the `myFunction` function.

   ```
   async function myFunction(Button button) {
     var text = await button.getText();
     // Do something with the passed text.
   }
   ```

   Listing 5.5 async Function

2. Now use the `then` function. In the example in Listing 5.6, you can see how to call the `getText` function of a button that was passed as a function argument to the `myFunction` function.

   ```
   function myFunction(Button button) {
     button.getText().then(myOtherFunction(text) {
       // Do something with the passed text.
     });
   }
   ```

   Listing 5.6 myFunction Function

3. If your browser doesn't know the keywords `async` and `await` or the `then` function, use a transpiler that offers the functionality of one of the patterns described above.

Importing Script API Data Types with Type Libraries

You can use script API data types when implementing custom widgets. These include, for example, properties, method arguments, and return values. Script API types are grouped together in *type libraries*.

[«]

> **Type Libraries**
>
> The use of type libraries is currently optional in the custom widget framework.

Importing type libraries

To use a specific script API data type in your web component, you must request the corresponding type library in the JSON file of the custom widget. In this example, you can see how to add the `input-controls` and `table` script API data types to the JSON file of the custom widget in order to use them in the web component:

```
"imports": ["table", "input-controls"]
```

Type Library: standard

You don't need to add the `standard` type library to the JSON file of the custom widget. This is loaded automatically.

You can find a list of all type libraries in the API Reference Guide at *http://s-prs.co/v594407*.

Use of Data Bindings

Data binding

You can use *data bindings* in a web component, but you must first define them in the JSON file of the custom widget. You can then access the data and call supported APIs.

In the example in Listing 5.7, you can see how the `myDataBinding` data binding is defined in the JSON file of the custom widget.

```
"dataBindings": {
  "myDataBinding": {
    "feeds": [
      {
        "id": "dimensions",
        "description": "Dimensions",
        "type": "dimension"
      },
      {
        "id": "measures",
        "description": "Measures",
        "type": "mainStructureMember"
      }
    ]
  }
}
```

Listing 5.7 myDataBinding

Accessing the data

Once you have stored the data binding in the JSON file, you can access the data via the name of the data binding:

```
this.myDataBinding
```

The following properties are available in myDataBinding, contain the following data, and perform the specified functions:

- The property data contains the rows of the result set.
- The feedName_index alias is used in the data rows.
- The property metadata contains the information required to run through the result set.

You can query result sets as shown in Listing 5.8.

```
// Traverse result set
this.myDataBinding.data.forEach(row => {
  // Parse row.
})
```

Listing 5.8 Parsing Result Set

Call data binding APIs

You can use the getDataBinding function to get a dataBindings object that contains data-driven APIs that the custom widget can execute. In the example in Listing 5.9, you can see how feed-related APIs are used to add a dimension.

```
const dataBinding = this.dataBindings.getDataBinding('myDataBinding')
await dataBinding.addDimensionToFeed("dimensions", dimensionId)
```

Listing 5.9 Example of getDataBinding

You can find a list of all APIs that you can call by a dataBindings object in the API Reference Guide at *http://s-prs.co/v594407*. Search there for the *Data-Binding* keyword.

5.2.3 Styling Panel

Features of the Styling Panel

The *styling panel* is an area in the story in which you define the properties of the custom widget, such as the color, when creating the story. The area is located on the **Styling** tab of the right-side panel and is titled **Custom Widgets Additional Properties.** Implementing the styling panel is optional, it's implemented as a web component, and it's contained in a separate JavaScript web component file.

[«]

Usable Functions of the Web Component

In the design time, the story creates a separate styling panel next to the canvas for each custom widget. Due to the fact that the styling panel is implemented as a web component, the same lifecycle functions and callbacks are available as can be called by the actual web component (with a value of `kind` = `main` in the JSON file). The `onCustomWidgetResize` function is an exception; it is never called.

[»]

Behavior When Removing the Custom Widget

If you delete a custom widget from the canvas of the design time, the styling panel is destroyed and the `disconnectedCallback` and `onCustomWidgetDestroy` lifecycle and callback functions are called. There is no defined sequence here. If you need to clear the status of the styling panel, we recommend doing this in the `onCustomWidgetDestroy` function.

5.2.4 Builder Panel

Features of the Builder Panel

The *builder panel* is an area in the story in which you define properties of the custom widget, such as measures and filters, when creating the story. Implementing the builder panel is optional. This area is located on the **Builder** tab of the right-side panel and is titled **Builder**. The builder panel takes up the entire screen area of the **Builder**, and it is implemented as a web component, which is contained in a separate JavaScript web component file.

[»]

Usable Functions of the Web Component

In the design time, the story creates a separate builder panel next to the canvas for each custom widget. Because the builder panel is implemented as a web component, the same lifecycle functions and callbacks that can be called by the actual web component (with a value of `kind` = `main` in the JSON file) are available. The `onCustomWidgetResize` function is an exception, as this function is never called.

[»]

Behavior When Removing the Custom Widget

If you delete a custom widget from the canvas of the design time, the builder panel is destroyed and the `disconnectedCallback` and `onCustomWidgetDestroy` lifecycle and callback functions are called. There is no defined sequence here. If you need to clear the status of the builder panel, we recommend that you do it in the `onCustomWidgetDestroy` function.

5.3 Creating, Uploading, and Removing a Custom Widget

In this section, we'll use a simple example to teach you how to create a custom widget, deploy it in your SAP Analytics cloud tenant, and remove it.

Infrastructural requirements

To host a custom widget, you need to either use a web server or host it directly on your SAP Analytics Cloud tenant. Both scenarios are presented in the following sections. You can find more detailed information on this and how you can use a local web server in Section 5.4.3.

5.3.1 Creation of a Custom Widget

Store files

To show you how to create a custom widget, we'll use a very simple example: a custom widget that is a colored box whose color and transparency you can define via the builder and styling panel.

The custom widget consists of a total of five files:

- *box.json*
- *box.js*
- *box_builder.js*
- *box_styling.js*
- *icon.png*

The custom widget we use here is described in in more detail in Section 5.4, and we also provide the source code there.

Create a new folder on your desktop and place the listed files there. In our example, this folder is called *coloredbox*, but if you use a different name, you may need to change the URLs to reference the resource files in the JSON file.

Figure 5.2 shows the files of the custom widget in the *Custom Widgets* parent folder.

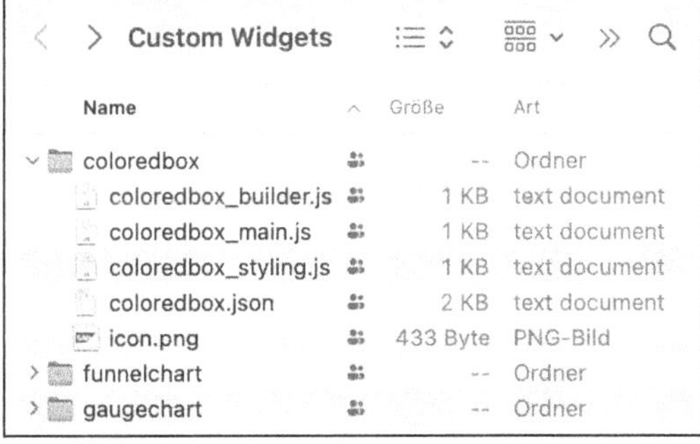

Figure 5.2 Folder of Custom Widget

5.3.2 Hosting a Custom Widget

Hosting custom widgets

As we already mentioned at the beginning of Section 5.2, the prerequisite for being able to use your own custom widget is that you have the infrastructure to make the resource file accessible on a web server. There are several options for this that differ in their approach but deliver the same result: hosting the resource files. Some options for this are as follows:

- You can operate your own web server and make the resource files accessible via it.
- You can switch to hyperscalers such as Amazon Web Services, Microsoft Azure, and Google Cloud Platform, and you can also use them to host static files.
- You can host the resource files via GitHub, which is a quick and easy solution, especially for development and testing purposes.
- You can host the resource files via your SAP Analytics Cloud tenant.
- You can use local hosting, which is another option for providing the static resource files. One advantage of this is that the files can't be accessed externally and remain within your firewall.

Deploying the local web server

In this example, we'll use a local web server to host the resource files. We'll use the *http-server* package from the *npm package manager*. With this package, files or folders from your desktop are made available in your network with little effort, but there are other providers on the market that can handle these tasks just as well.

Starting the local web server

There are several ways to install npm packages, and one way is to install them via the *Node.js* JavaScript runtime environment. To do this, you must first download and install *Node.js* via *https://nodejs.org*, and then, you'll be able to access the npm library. In the description of the http-server package, which you can access at via *https://www.npmjs.com/package/http-server*, you'll find various options for installation. If you use *npx* as tool, you can start the script (i.e., the web server) without installation.

Execute this command via the terminal or via the console if you are in the folder that also contains the *coloredbox* folder:

```
npx http-server
```

As shown in Figure 5.3, server properties are logged when the server is started. If the start is successful, the internet protocol address and port are displayed. In this figure, the server can be reached at *http://127.0.0.1:8080*. You can also test this directly.

Figure 5.4 shows you how the files are exposed via the web server. It indicates that there are three directories: *coloredbox*, *funnelchart*, and *gaugechart*.

```
Custom Widgets — http-server ◂ npm exec h...
                   Custom Widgets % npx http-server
Starting up http-server, serving ./

http-server version: 14.1.1

http-server settings:
CORS: disabled
Cache: 3600 seconds
Connection Timeout: 120 seconds
Directory Listings: visible
AutoIndex: visible
Serve GZIP Files: false
Serve Brotli Files: false
Default File Extension: none

Available on:
  http://127.0.0.1:8080
  http://10.181.219.142:8080
Hit CTRL-C to stop the server
```

Figure 5.3 Start of HTTP Server

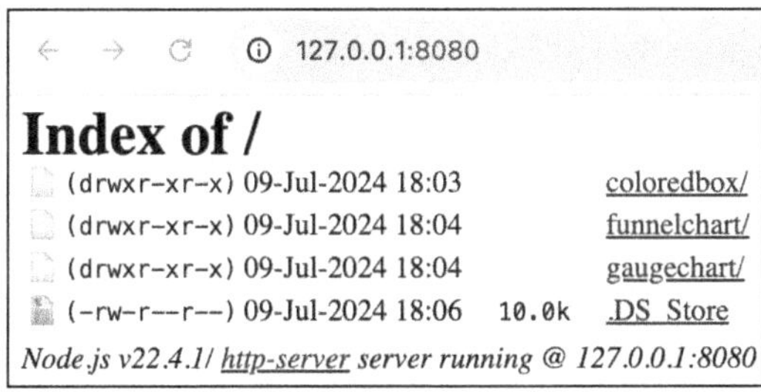

Figure 5.4 http-server Endpoint

Opening custom widget resource files via the web server

You can click on *coloredbox/* to open the directory, and the resource files of the custom widget will be displayed in a list (see Figure 5.5). If you click on one of the files, you can display the source code of that file.

```
127.0.0.1:8080/coloredbox/

Index of /coloredbox/
(drwxr-xr-x) 09-Jul-2024 18:04         ../
(-rw-r--r--) 09-Jul-2024 17:31  6.0k   .DS_Store
(-rw-r--r--) 09-Jul-2024 17:17  1.2k   coloredbox_builder.js
(-rw-r--r--) 09-Jul-2024 17:16  1.0k   coloredbox_main.js
(-rw-r--r--) 09-Jul-2024 17:18  1.1k   coloredbox_styling.js
(-rw-r--r--) 09-Jul-2024 17:56  1.6k   coloredbox.json
(-rw-r--r--) 09-Jul-2024 17:30  433B   icon.png
Node.js v22.4.1/ http-server server running @ 127.0.0.1:8080
```

Figure 5.5 coloredbox Directory

Now the custom widget is successfully hosted on a web server, and you can access it from your desktop.

5.3.3 Uploading a Custom Widget

Adding custom widgets in SAP Analytics Cloud

In this section, we'll show you how to upload a custom widget to SAP Analytics Cloud—though the word *upload* isn't quite accurate here since the resource files remain your responsibility and are made available via the web server. Only the JSON file of the custom widget is uploaded.

Follow these steps to upload the custom widget:

1. Open **Stories** via the navigation bar and switch to the **Custom Widgets** tab (see Figure 5.6).

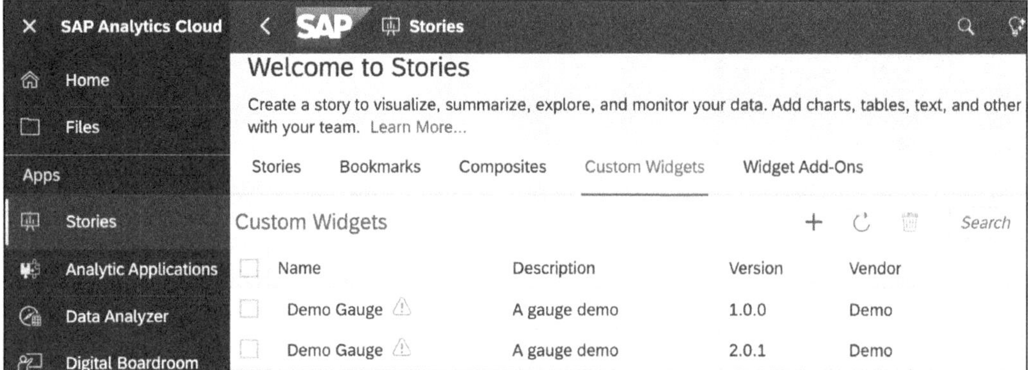

Figure 5.6 Managing Custom Widgets

2. Click the + icon, which will cause the **Upload File** popup to open.
3. Click the **Select File** button if the popup for selecting files in the file system doesn't open automatically (see Figure 5.7).

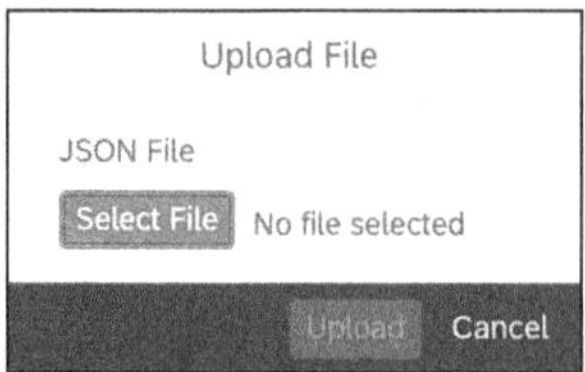

Figure 5.7 Popup Dialog for Selecting JSON File

4. Select a JSON file (in this example, *coloredbox.json*).

If you upload *coloredbox.json* as we provide it, a note should be displayed after you select the file (see Figure 5.8). It refers to the fact that the custom widget is currently still in so-called *development mode*, which is due to the fact that sub resource integrity checks are suppressed in the JSON file. This isn't a problem for development and testing purposes, but for productive use of the custom widget, we recommend activating the integrity check.

Section 5.4.1 takes a closer look at the JSON document and will show you how to enable and disable the integrity check.

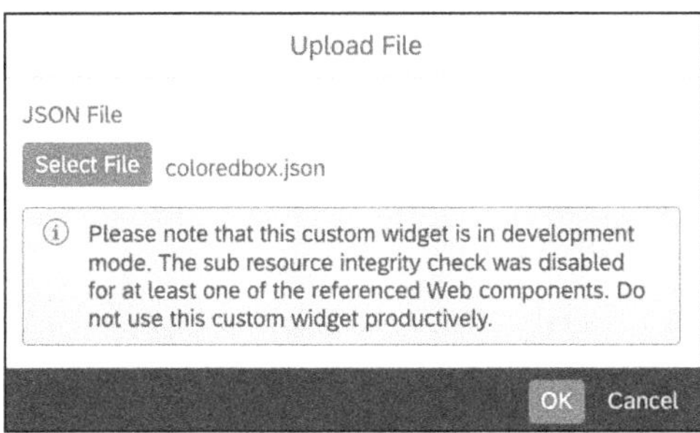

Figure 5.8 Note About Development Mode

After the JSON file has been uploaded, a new custom widget will be listed in the **Custom Widgets** tab. As the custom widget is still in development mode, it has a yellow warning flag in this overview (see Figure 5.9).

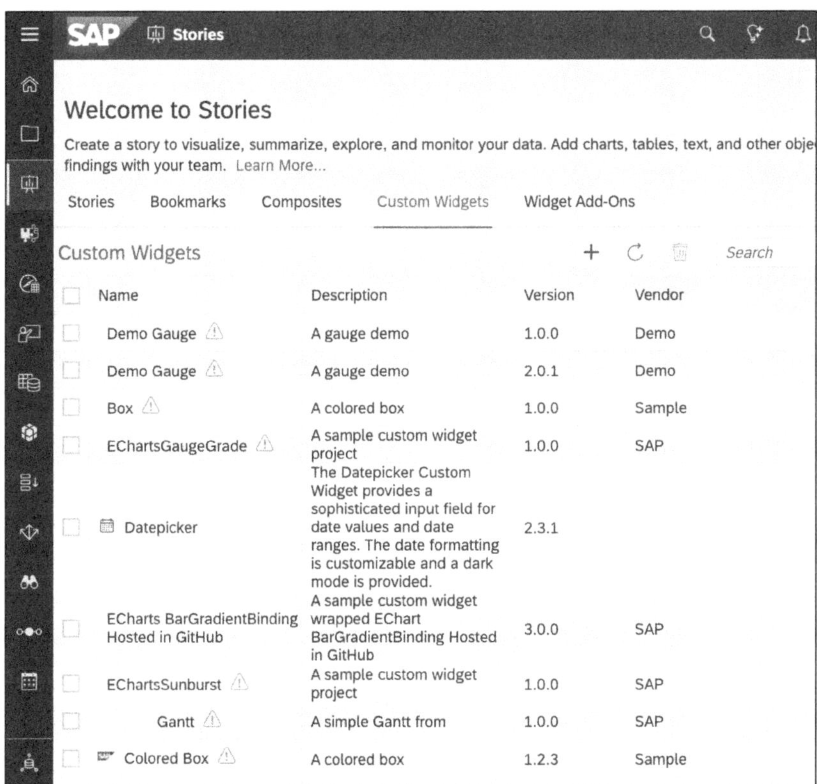

Figure 5.9 List of All Custom Widgets Currently Uploaded

Adding custom widgets to a story

Once you have uploaded the custom widget, you can use it in a story. To do this, open a story and the left-side panel. Under the **Assets** tab, you'll find the **Custom Widgets** section, where all available custom widgets are listed (see Figure 5.10). You can drag and drop the widget you want into your story.

Please note that you need special permission to upload and update custom widgets.

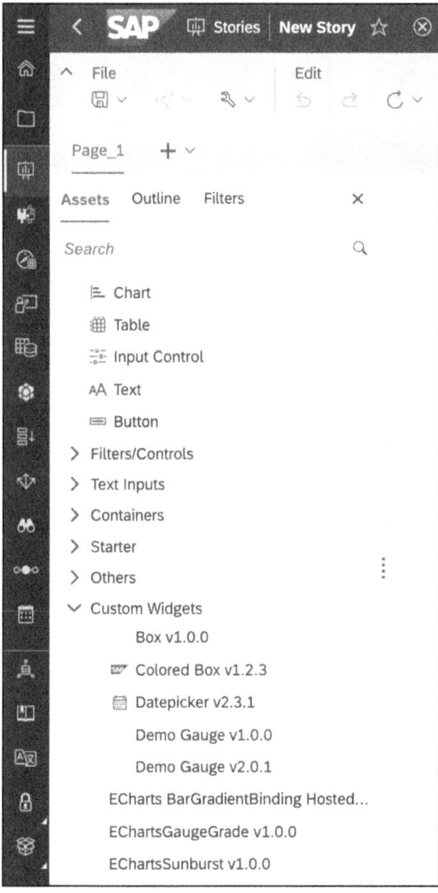

Figure 5.10 Adding Custom Widgets to Story

Versioning Custom widgets have a version number, and Figure 5.9 shows you the format in which it is maintained. In our example, the custom widget has the version number 1.2.3. Each of the numbers stands for one of three dimensions: *majorVersion*, *minorVersion*, and *patchVersion*. The version status is also defined via the reference document, which is the JSON file. In SAP Analytics Cloud, you can upload custom widgets with different versions, so it's important that the version status of a custom widget differs from the major version. If a custom widget with the same major version is uploaded, the existing widget is always overwritten—regardless of the minor version. For example, if you currently have a custom widget of version 2.2.0 in SAP Analytics Cloud and you upload a widget of version 2.1.0 or 2.3.0, the existing version 2.2.0 is overwritten regardless of the version of the widget to be uploaded.

5.3.4 Deleting Custom Widgets

Deleting a custom widget

You can delete custom widgets from your SAP Analytics Cloud Tenant in just a few steps:

1. Open **Stories** via the navigation bar and switch to the **Custom Widgets** tab (refer back to Figure 5.6).
2. Select the custom widget you want to delete.
3. Click the bin icon to delete it.
4. In the popup box that appears after you click the bin icon, click the **Delete** button to confirm the deletion.

5.4 An Example of a Custom Widget: The Colored Box

Functions of the colored box

In this section, we'll guide you through the code of a simple custom widget, the *colored box*. This custom widget is a red box with a black border, but this example covers all the topics that are relevant to custom widgets. The main points about a custom widget that the example highlights are as follows:

- It can be added to the canvas.
- It can be moved and resized.
- It adds script methods to the script editor.
- It adds a builder and a styling panel in which the properties of the custom widget can be changed in the design time.

The custom widget is made up of three web components: the actual box, the builder panel, and the styling panel. The example widget consists of the files in Table 5.14.

File	Description
coloredbox.json	This is the JSON file of the custom widget. You can find the source code in Section 5.4.1.
coloredbox_main.js	This is the JavaScript web component of the colored box. You can find the source code in Section 5.4.2.

Table 5.14 Overview of Files Used for Sample Custom Widget

File	Description
coloredbox_styling.js	This is the JavaScript web component of the styling panel of the colored box. You can find the source code in Section 5.4.3.
coloredbox_builder.js	This is the JavaScript web component of the builder panel of the colored box. You can find the source code i n Section 5.4.4.
icon.png	This is the icon of the colored box, with a resolution of 16 × 16 pixels.

Table 5.14 Overview of Files Used for Sample Custom Widget (Cont.)

In Figure 5.11, you can see what the custom widget looks like, and you can also see that the styling panel is open. Additional properties are available for this custom widget; for example, you can define the color of the box in the **Colored Box Properties** section.

Figure 5.11 Custom Widget with Styling Panel

Figure 5.12 shows you the custom widget again, but this time, with the builder panel open. Additional setting options are also available under **Colored Box Properties**, which lets you set how transparent the color of the custom widget should be.

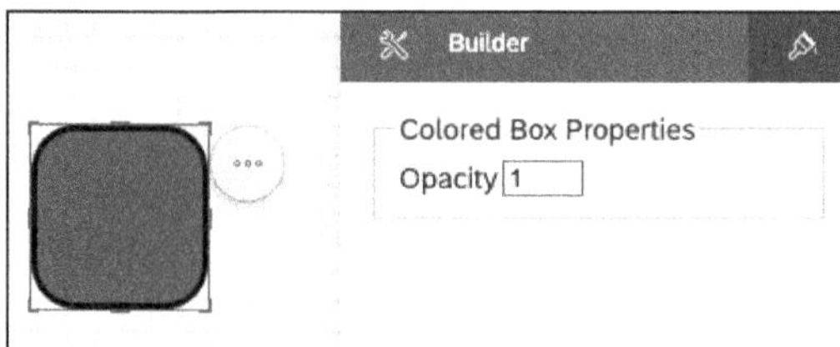

Figure 5.12 Custom Widget with Builder Panel

5.4.1 JSON File of the Custom Widget

So that you can understand our example custom widget, Listing 5.10 shows the source code of the *coloredbox.json* JSON file of the custom widget.

```
{
    "id": "com.sap.sample.coloredbox",
    "version": "1.2.3",
    "name": "Colored Box",
    "description": "A colored box",
    "newInstancePrefix": "ColoredBox",
    "icon": "http://127.0.0.1:8080/coloredbox/icon.png",
    "vendor": "Sample",
    "eula": "",
    "license": "",
    "webcomponents": [
        {
            "kind": "main",
            "tag": "com-sap-sample-coloredbox",
            "url": "http://127.0.0.1:8080/coloredbox/coloredbox_
main.js",
            "integrity": "",
            "ignoreIntegrity": true
        },
        {
            "kind": "styling",
            "tag": "com-sap-sample-coloredbox-styling",
            "url": "http://127.0.0.1:8080/coloredbox/coloredbox_
styling.js",
            "integrity": "",
            "ignoreIntegrity": true
        },
        {
            "kind": "builder",
            "tag": "com-sap-sample-coloredbox-builder",
            "url": "http://127.0.0.1:8080/coloredbox/coloredbox_
builder.js",
            "integrity": "",
            "ignoreIntegrity": true
        }
    ],
    "properties": {
        "color": {
            "type": "string",
            "description": "Background color",
```

```
            "default": "red"
        },
        "opacity": {
            "type": "number",
            "description": "Opacity",
            "default": 1
        },
        "width": {
            "type": "integer",
            "default": 100
        },
        "height": {
            "type": "integer",
            "default": 100
        }
    },
    "methods": {
        "setColor": {
            "description": "Sets the background color.",
            "parameters": [
                {
                    "name": "newColor",
                    "type": "string",
                    "description": "The new background color"
                }
            ],
            "body": "this.color = newColor;"
        },
        "getColor": {
            "returnType": "string",
            "description": "Returns the background color.",
            "body": "return this.color;"
        }
    },
    "events": {
        "onClick": {
            "description": "Called when the user clicks the Colored
Box."
        }
    }
}
```

Listing 5.10 Source Code of coloredbox.json

In Listing 5.11, we'll now look at the individual sections of the JSON file of the custom widget. We start with the top section:

```
"id": "com.sap.sample.coloredbox",
"version": "1.2.3",
"name": "Colored Box",
"description": "A colored box",
"newInstancePrefix": "ColoredBox",
"icon": "http://127.0.0.1:8080/coloredbox/icon.png",
"vendor": "Sample",
"eula": "",
"license": "",
```

Listing 5.11 First Section of coloredbox.json

The custom widget has the unique ID `com.sap.sample.coloredbox`, is version 1.2.3, and has the name *Colored Box*, which is displayed in SAP Analytics Cloud. The custom widget also has the description *A colored box*, which is used in the tooltips, for example. The prefix of the new instance of the custom widget is *ColoredBox*, the icon for the colored box is provided via a URL, and the provider is *Sample*. The end user agreement and the license field have been left blank.

The custom widget colored box consists of the three web components in Listing 5.12.

```
"webcomponents": [
    {
        "kind": "main",
        "tag": "com-sap-sample-coloredbox",
        "url": "http://127.0.0.1:8080/coloredbox/coloredbox_
main.js",
        "integrity": "",
        "ignoreIntegrity": true
    },
    {
        "kind": "styling",
        "tag": "com-sap-sample-coloredbox-styling",
        "url": "http://127.0.0.1:8080/coloredbox/coloredbox_
styling.js",
        "integrity": "",
        "ignoreIntegrity": true
    },
    {
        "kind": "builder",
        "tag": "com-sap-sample-coloredbox-builder",
```

```
            "url": "http://127.0.0.1:8080/coloredbox/coloredbox_
builder.js",
            "integrity": "",
            "ignoreIntegrity": true
        }
    ],
```

Listing 5.12 Definition of Custom Widget's Web Components

Web components The first web component is the colored box, its name is `com-sap-sample-coloredbox`, and it's classified by the `main` attribute. The second web component is the styling panel of the custom widget, the component's name is `com-sap-sample-box-styling`, and it's defined by the `styling` attribute. The third web component is the builder panel of the custom widget, the component's name is `com-sap-sample-box-builder`, and it's defined by the `builder` attribute. The JavaScript file of each web component is referred to via a URL. In these examples, no hash value was added to check the integrity of the referenced files. For all web components, the `ignoreIntegrity` property was set to `true`, and thus, the integrity check was suspended.

Next, let's look at the properties of the custom widget in Listing 5.13.

```
    "properties": {
        "color": {
            "type": "string",
            "description": "Background color",
            "default": "red"
        },
        "opacity": {
            "type": "number",
            "description": "Opacity",
            "default": 1
        },
        "width": {
            "type": "integer",
            "default": 100
        },
        "height": {
            "type": "integer",
            "default": 100
        }
    },
```

Listing 5.13 Definition of Custom Widget's Properties

Properties and their significance

The color property specifies the color of the box, it's of the String type, and the default value is red. The opacity property specifies the opacity or transparency of the box, it's of the Number type, and it has a default value of 1. The value can be between 0 and 1, where 1 means that the custom widget completely covers the background.

The width and height properties specify the initial height and width of the box. Both have the value integer with a default value of 100 pixels, and both are also special properties in the custom widget framework. Implemented setter functions are not called for these properties.

In the next section of the JSON file, the script methods are defined as shown in Listing 5.14.

```
"methods": {
    "setColor": {
        "description": "Sets the background color.",
        "parameters": [
            {
                "name": "newColor",
                "type": "string",
                "description": "The new background color"
            }
        ],
        "body": "this.color = newColor;"
    },
    "getColor": {
        "returnType": "string",
        "description": "Returns the background color.",
        "body": "return this.color;"
    }
},
```

Listing 5.14 Definition of Custom Widget's Script Methods

Two script methods are defined: setColor and getColor. The setColor function has one argument: the new color. The body property of the function takes the argument and sets it as the new color in the color property of the custom widget. The getColor function has no arguments. Instead, the body property contains a script that returns a value as a string: the current color of the box.

Finally, an onClick event is defined in the JSON document (see Listing 5.15).

```
"events": {
    "onClick": {
        "description": "Called when the user clicks the Colored
```

```
Box."
            }
    }
```

Listing 5.15 Definition of Custom Widget's Event

5.4.2 JavaScript Web Component

The code in Listing 5.16 shows the source code of the *coloredbox_main.js* JavaScript file of the web component of the custom widget. The code is described in more detail in the following sections.

```
(function() {
let template = document.createElement("template");
template.innerHTML = `
<style>
:host {
border-radius: 25px;
border-width: 4px;
border-color: black;
border-style: solid;
display: block;
}
</style>
`;

class ColoredBox extends HTMLElement {
constructor() {
super();
let shadowRoot = this.attachShadow({mode: "open"});
shadowRoot.appendChild(template.content.cloneNode(true));
this.addEventListener("click", event => {
var event = new Event("onClick");
this.dispatchEvent(event);
});
this._props = {};
}

onCustomWidgetBeforeUpdate(changedProperties) {
this._props = { ...this._props, ...changedProperties };
}

onCustomWidgetAfterUpdate(changedProperties) {
if ("color" in changedProperties) {
this.style["background-color"] = changedProperties["color"];
}
```

```
if ("opacity" in changedProperties) {
this.style["opacity"] = changedProperties["opacity"];
}
}
}

customElements.define("com-sap-sample-coloredbox", ColoredBox);
})();
```

Listing 5.16 Source Code of coloredbox_main.js

Definition of the Custom Element and Its JavaScript API

The web component defines a new `com-sap-sample-coloredbox` custom element and links it to a JavaScript API. In our example, this is represented by the `ColoredBox` class (see Listing 5.17).

```
(function() {
...
class ColoredBox extends HTMLElement {
...
}

customElements.define("com-sap-sample-coloredbox", ColoredBox);
})();
```

Listing 5.17 Definition of New Custom Element

Structure of the implementation

The entire JavaScript code of the file is nested in a self-executing JavaScript function, which isolates the scope of its variables from the global scope. The JavaScript API of the new custom element is implemented in the `ColoredBox` class, which extends the JavaScript API of the `HTMLElement` class. Calling `customElements.define()` defines a new custom element named `com-sap-sample-coloredbox` and links it to the JavaScript API represented by the `ColoredBox` class. Note that `com-sap-sample-coloredbox` is the same name as the property value of the tag of the property name of the first element in the array of web components in the JSON file of the custom widget.

Creating a Template Object

The next section of code creates a template of an HTML element (see Listing 5.18).

```
let template = document.createElement("template");
template.innerHTML = `
<style>
:host {
```

```
border-radius: 25px;
border-width: 4px;
border-color: black;
border-style: solid;
display: block;
}
</style>
`;
```

Listing 5.18 Definition of HTML Template

Advantages of the HTML template

Template objects are important parts of web components. The example in Listing 5.18 creates a template HTML object that serves as a template for the shadow HTML object. This represents the HTML DOM of the box, which exists separately from the HTML DOM of the website. It's only added to the HTML DOM of the website when the custom widget is rendered. This separation allows you to apply changes to the shadow DOM element, for example, without having to take the HTML DOM of the website into account.

The template from the example widget is very simple. It defines a CSS style from several CSS properties together with the `:host` CSS selector. This CSS style is assigned to the template object, and the `:host` selector selects the root of the shadow DOM element (i.e., the root of the template). The `display` CSS property is used to configure the display of the custom HTML element, and possible values include `block`, `flex`, and `inline-block`.

Definition of the JavaScript API of the Custom Element

In the next section, the JavaScript API of the custom element is defined by the `ColoredBox` class. This extends the `HTMLElement` class as follows:

```
class ColoredBox extends HTMLElement {
...
}
```

Implementation of the Constructor Function

As you can see in the code of the JavaScript file, the first function defined by the JavaScript API is the `constructor()` function (see Listing 5.19).

```
constructor() {
super();
let shadowRoot = this.attachShadow({mode: "open"});
shadowRoot.appendChild(template.content.cloneNode(true));
this.addEventListener("click", event => {
var event = new Event("onClick");
this.dispatchEvent(event);
```

```
});
this._props = {};
}
```

Listing 5.19 Definition of constructor() Function

How the constructor() function works

First, the super() function is called, and then, the shadow DOM root element is created and a copy of the template object is added to the shadow DOM root element as a subordinate element. Finally, an event listener is added to the custom element. This waits for click events and triggers an onClick event as soon as one occurs. Pay attention to the correct naming. The name of the onClick event must be the same as in the events section of the JSON file of the custom widget.

An empty _props object is initialized to make it easier to manage the properties of the web component.

Dealing with Updates to the Custom Widget

By implementing the onCustomWidgetBeforeUpdate and onCustomWidgetAfterUpdate functions, the colored box processes the updates to its color and opacity properties. These are defined in the properties array of the JSON file.

The onCustomWidgetBeforeUpdate function merges the properties of the changedProperties object with those of the _props object. This means that _props contains all the properties of the box before it's updated, as you can see here:

```
onCustomWidgetBeforeUpdate(changedProperties) {
this._props = { ...this._props, ...changedProperties };
}
```

The onCustomWidgetAfterUpdate function uses the transferred properties of the changedProperties object to adjust the color and opacity CSS styles of the custom widget (see Listing 5.20).

```
onCustomWidgetAfterUpdate(changedProperties) {
if ("color" in changedProperties) {
        this.style["background-color"] = changedProperties["color"];
    }
    if ("opacity" in changedProperties) {
        this.style["opacity"] = changedProperties["opacity"];
    }
}
```

Listing 5.20 Definition of onCustomWidgetAfterUpdate Function

Implementation of Property Setter Functions

Another way to implement the updating of the `color` and `opacity` properties of the box is to use property setter functions.

Use of Property Setter Functions

Using this approach only makes sense for simple custom widgets with a few independent properties. The more properties the custom widget has, the more likely it is that (complex) dependencies will arise between the properties. With multiple properties, managing the dependencies by implementing an `onCustomWidgetAfterUpdate` function is the preferred approach.

The code in Listing 5.21 shows you the implementation of a property setter function.

```
set color(newColor) {
this.style["background-color"] = newColor;
}
set opacity(newColor) {
this.style["opacity"] = newColor;
}
```

Listing 5.21 Definition of Setter Functions for Color and Opacity

Property setter functions carry the name of the property defined in the JSON file of the custom widget. According to the object accessor notation, a setter function uses the keyword `set`.

The implementation is simple. In this example, a setter function of the `color` property is assigned the transferred value of the new color of the `background-color` CSS property of the custom element.

5.4.3 Styling Panel JavaScript Web Component

The code in Listing 5.22 shows you the source code of the *coloredbox_styling.js* JavaScript file of the styling panel of the custom widget. You can change the color of the colored box in the styling panel of the story. The source code is explained step by step below, and it is very similar to the JavaScript file of the web component of the box.

```
(function() {
    let template = document.createElement("template");
    template.innerHTML = `
        <form id="form">
```

```
            <fieldset>
                <legend>Colored Box Properties</legend>
                <table>
                    <tr>
                        <td>Color</td>
                        <td><input id="styling_color" type="text"
size="40" maxlength="40"></td>
                    </tr>
                </table>
                <input type="submit" style="display:none;">
            </fieldset>
        </form>
    `;

    class ColoredBoxStylingPanel extends HTMLElement {
        constructor() {
            super();
            this._shadowRoot = this.attachShadow({mode: "open"});
            this._
shadowRoot.appendChild(template.content.cloneNode(true));
            this._
shadowRoot.getElementById("form").addEventListener("submit", this._
submit.bind(this));
        }

        _submit(e) {
            e.preventDefault();
            this.dispatchEvent(new CustomEvent("propertiesChanged", {
                    detail: {
                        properties: {
                            color: this.color
                        }
                    }
            }));
        }

        set color(newColor) {
            this._shadowRoot.getElementById("styling_color").value =
newColor;
        }

        get color() {
            return this._shadowRoot.getElementById("styling_
color").value;
```

```
        }
    }

customElements.define("com-sap-sample-coloredbox-styling",
ColoredBoxStylingPanel);
})();
```

Listing 5.22 Source Code of coloredbox_styling.js

Definition of the Custom Element and Its JavaScript API

The web component defines a new `com-sap-sample-coloredbox-styling` custom element and links it to a JavaScript API. This is represented in the example by the `ColoredBoxStylingPanel` class, as shown in Listing 5.23.

```
(function() {
...
class ColoredBoxStylingPanel extends HTMLElement {
...
}

customElements.define("com-sap-sample-coloredbox-styling",
ColoredBoxStylingPanel);
})();
```

Listing 5.23 Definition of New Custom Element

The structure of the implementation

The entire JavaScript code of the file is nested in a self-executing JavaScript function, which isolates the scope of its variables from the global scope. The JavaScript API of the new custom element is implemented in the `ColoredBoxStylingPanel` class, which extends the JavaScript API of the `HTMLElement` class. The `customElements.define()` call defines a new custom element with the name `com-sap-sample-coloredbox-styling` and links it to the JavaScript API represented by the `ColoredBoxStylingPanel` class.

Creating a Template Object

The code shown in Listing 5.24 creates a template of an HTML element.

```
let template = document.createElement("template");
template.innerHTML = `
    <form id="form">
        <fieldset>
            <legend>Colored Box Properties</legend>
            <table>
                <tr>
                    <td>Color</td>
```

```
                    <td><input id="styling_color" type="text" size=
"40" maxlength="40"></td>
                  </tr>
               </table>
               <input type="submit" style="display:none;">
            </fieldset>
        </form>
`;
```

Listing 5.24 Definition of HTML Template

This example creates a template HTML object that serves as a template for the shadow HTML object. This represents the HTML DOM of the format area of the box.

This template object defines a simple form with the ID `form`. This form contains a table with an input field, the ID `styling_color`, and a maximum length of forty characters.

[+]

Defining a Hidden Input Field

Defining a hidden input field of the `submit` type is a trick to make the form work if the form contains more than one input element. However, this is only necessary when you add more input elements.

Definition of the JavaScript API of the Custom Element

The JavaScript API of the custom element is defined by the `ColoredBoxStylingPanel` class. This extends the `HTMLElement` class:

```
class ColoredBoxStylingPanel extends HTMLElement {
...
}
```

Listing 5.25 Definition of ColoredBoxStylingPanel Class

Implementation of the Constructor Function

The first function defined by the JavaScript API is the `constructor()` function, as shown in Listing 5.26.

```
constructor() {
super();
this._shadowRoot = this.attachShadow({mode: "open"});
this._shadowRoot.appendChild(template.content.cloneNode(true));
```

```
this._shadowRoot.getElementById("form").addEventListener("submit",
this._submit.bind(this));
}
```

Listing 5.26 Definition of constructor() Function

How the constructor() function works

First, the super() function is called, and then, the shadow DOM root element is created. A reference to the root element is stored in the _shadowRoot local variable, as the shadow DOM root element is referenced later in the setter and getter functions. Then, a copy of the template object is added to the shadow DOM root element as a subordinate element, and finally, an event listener is added to the custom element. This waits for submit events and triggers the _submit() event handler function as soon as one occurs. Calling bind() and passing this to _submit() ensures that the keyword this in _submit() refers to the custom element.

The _submit() function is implemented as shown in Listing 5.27.

```
_submit(e) {
    e.preventDefault();
    this.dispatchEvent(new CustomEvent("propertiesChanged", {
        detail: {
            properties: {
                color: this.color
            }
        }
    }));
}
```

Listing 5.27 Definition of _submit() Function

How the _submit() function works

The _submit() function calls the preventDefault() function for the passed event object. This prevents the form from being transferred to the server.

A custom event with the name propertiesChanged is then created, and it contains JSON information that defines which properties have changed and which new property values they take on. In this example, it's the color property of the custom widget, and its new property value is retrieved by executing this.color, which implicitly calls the getter function of the color property of the ColoredBoxStylingPanel class.

Implementation of Property Setter and Property Getter Functions

Listing 5.28 shows you the implementation of the property setter and property getter functions.

[«]

Risk of Confusion

This is a different function than the setter function of the color property of the ColoredBox class.


```
set color(newColor) {
this._shadowRoot.getElementById("styling_color").value = newColor;
}

get color() {
return this._shadowRoot.getElementById("styling_color").value;
}
```

Listing 5.28 Definition of Setter and Getter Functions for Color

How set color works

The setter function sets the text representation of the new color in the input field of the styling panel of the custom widget. This function isn't called directly; the custom widget framework calls it when the styling panel of the custom widget is created and assigns an initial value to the input field.

How get color works

The getter function returns the text of the input field of the styling panel of the custom widget. This function is called indirectly by the _submit() function of the event handler during the creation of the JSON information when the current color is retrieved by this.color.

5.4.4 Builder Panel JavaScript Web Component

The code in Listing 5.29 shows you the source code of the *coloredbox_builder.js* JavaScript file of the builder panel of the custom widget. You can change the opacity of the colored box in the builder panel of the story.

In the following sections, the source code in Listing 5.29 is explained step by step. This is very similar to the JavaScript file of the web component of the styling panel of the custom widget.

```
(function() {
    let template = document.createElement("template");
    template.innerHTML = `
        <form id="form">
            <fieldset>
                <legend>Colored Box Properties</legend>
                <table>
                    <tr>
                        <td>Opacity</td>
```

```
                        <td><input id="builder_opacity" type="text"
size="5" maxlength="5"></td>
                    </tr>
                </table>
                <input type="submit" style="display:none;">
            </fieldset>
        </form>
        <style>
        :host {
            display: block;
            padding: 1em 1em 1em 1em;
        }
        </style>
    `;

    class ColoredBoxBuilderPanel extends HTMLElement {
        constructor() {
            super();
            this._shadowRoot = this.attachShadow({mode: "open"});
            this._
shadowRoot.appendChild(template.content.cloneNode(true));
            this._
shadowRoot.getElementById("form").addEventListener("submit", this._
submit.bind(this));
        }

        _submit(e) {
            e.preventDefault();
            this.dispatchEvent(new CustomEvent("propertiesChanged", {
                    detail: {
                        properties: {
                            opacity: this.opacity
                        }
                    }
            }));
        }

        set opacity(newOpacity) {
            this._shadowRoot.getElementById("builder_opacity").value
= newOpacity;
        }

        get opacity() {
            return this._shadowRoot.getElementById("builder_
```

```
opacity").value;
        }
    }

    customElements.define("com-sap-sample-coloredbox-builder",
ColoredBoxBuilderPanel);
})();
```

Listing 5.29 Source Code of coloredbox_builder.js

Definition of the Custom Element and Its JavaScript API

The web component defines a new `com-sap-sample-coloredbox-builder` custom element and links it to a JavaScript API. This is represented in the example by the `ColoredBoxBuilderPanel` class (see Listing 5.30).

```
(function() {
...
class ColoredBoxBuilderPanel extends HTMLElement {
...
}

customElements.define("com-sap-sample-coloredbox-builder",
ColoredBoxBuilderPanel);
})();
```

Listing 5.30 Definition of New Custom Element

The structure of the implementation

The entire JavaScript code of the file is nested in a self-executing JavaScript function, which isolates the scope of its variables from the global scope. The JavaScript API of the new custom element is implemented in the `ColoredBoxBuilderPanel` class, which extends the JavaScript API of the `HTMLElement` class. The `customElements.define()` call defines a new custom element with the name `com-sap-sample-coloredbox-builder` and links it to the JavaScript API represented by the `ColoredBoxBuilderPanel` class.

Creating a Template Object

The next section of code, which you can find in Listing 5.31, creates a template of an HTML element.

```
let template = document.createElement("template");
template.innerHTML = `
    <form id="form">
        <fieldset>
            <legend>Colored Box Properties</legend>
            <table>
```

```
                <tr>
                    <td>Opacity</td>
                    <td><input id="builder_opacity" type="text" size=
"5" maxlength="5"></td>
                </tr>
            </table>
            <input type="submit" style="display:none;">
        </fieldset>
    </form>
    <style>
    :host {
        display: block;
        padding: 1em 1em 1em 1em;
    }
    </style>
`;
```

Listing 5.31 Definition of HTML Element

The Listing 5.31 creates a template HTML object that serves as a template for the shadow HTML DOM object. This represents the HTML DOM of the builder panel of the custom widget.

This template object defines a simple form with the ID `form`. This form contains a table with an input field, the ID `builder_opacity`, and a maximum length of five characters.

Defining a Hidden Input Field

Defining a hidden input field of the `submit` type is a trick to make the form work if the form contains more than one input element. This is still necessary now, but you should do it at the latest when you add more input elements.

The `:host` selector selects the root of the shadow DOM element (i.e., the root of the template). The `display` CSS property is used to configure the display of the custom HTML element (in this case, as a `block`), and the `padding` CSS property ensures that the form has sufficient space.

Definition of the JavaScript API of the Custom Element

The JavaScript API of the custom element is defined by the `ColoredBoxBuilderPanel`, which extends the `HTMLElement` class:

```
class ColoredBoxBuilderPanel extends HTMLElement {
...
}
```

Implementation of the Constructor Function

The first function defined by the JavaScript API is the `constructor()` function (see Listing 5.32).

```
constructor() {
super();
    this._shadowRoot = this.attachShadow({mode: "open"});
    this._shadowRoot.appendChild(template.content.cloneNode(true));
this._shadowRoot.getElementById("form").addEventListener("submit",
this._submit.bind(this));
}
```

Listing 5.32 Definition of constructor() Function

How the constructor() function works

First, the `super()` function is called, as you already know from the JavaScript API in the styling panel. The shadow DOM root element is then created, and a reference to it is saved in the `_shadowRoot` local variable, as the shadow DOM root element is referenced later in the setter and getter functions. Then, a copy of the template object is added to the shadow DOM root element as a subordinate element, and finally, an event listener is added to the custom element. This waits for `submit` events and triggers the `_submit()` event handler function as soon as one occurs. By calling `bind()` and passing `this` to `_submit()`, it's ensured that the keyword `this` in `_submit()` refers to the custom element.

The `_submit()` function is also implemented here as shown in Listing 5.33.

```
_submit(e) {
e.preventDefault();
    this.dispatchEvent(new CustomEvent("propertiesChanged", {
        detail: {
            properties: {
                opacity: this.opacity
            }
        }
    }));
}
```

Listing 5.33 Definition of _submit() Function

How the _submit() function works

The `_submit()` function calls the `preventDefault()` function for the passed event object. This prevents the form from being transferred to the server. A custom event with the name `propertiesChanged` is then created, and it contains JSON information that defines which properties have changed and which new property values they take on. In this example, the `opacity` property of the custom widget has changed, and its new property value is retrieved by executing `this.opacity`, which implicitly calls the getter function of the `opacity` property of the `ColoredBoxBuilderPanel` class.

Implementation of Property Setter and Property Getter Functions

The code in Listing 5.34 shows you the implementation of property setter and property getter functions.

```
set opacity(newOpacity) {
this._shadowRoot.getElementById("builder_opacity").value =
newOpacity;
}

get opacity() {
return this._shadowRoot.getElementById("builder_opacity").value;
}
```

Listing 5.34 Definition of Setter and Getter Functions for Opacity

How the setter function works

The setter function sets the text representation of the new opacity in the input field of the builder panel of the custom widget. This function isn't called directly, but by the custom widget framework when the builder panel of the custom widget is created and assigns an initial value to the input field.

How the getter function works

The getter function returns the text of the input field of the builder panel of the custom widget. This function is called indirectly by the `_submit()` function of the event handler during the creation of the JSON information when the current opacity is retrieved by `this.opacity`.

5.5 Custom Widgets on Mobile Devices

In this section, we explain what you need to consider if you want to use custom widgets on mobile devices and what special features you need to bear in mind when developing the widgets.

5.5.1 Activation of Custom Widgets for Mobile Devices

To activate custom widgets for mobile devices, you must set the `supportsMobile` property to `true`. You can find further information on this property in Section 5.4.1.

supportsMobile

[!]

> **Check Before Activation**
>
> Check whether your custom widget works properly on mobile devices before you activate this property.

5.5.2 Recognizing the End Device

Using the mobileMode property

You can recognize at runtime whether the custom widget is currently being used on a mobile device. If you have activated the `supportsMobile` property, the `mobileMode` property in the `otherProperties` array is passed to the `onCustomWidgetBeforeUpdate` and `onCustomWidgetAfterUpdate` functions. You can find further information on this in the next section.

5.5.3 Useful Tips for Developing Custom Widgets for Mobile Devices

If you are developing a custom widget for a mobile device, you should follow a few recommendations:

Simplifying interactions

- Enlarge the areas that that users will interact with. Areas that allow interaction in your custom widget should be large enough for easy use, so for example, you should enlarge clickable elements when the custom widget is running on a mobile device. This will allow users to tap the elements more easily with their fingers.

Alternatives for complex inputs

- Offer alternatives to function keys. Users can press function keys like `Ctrl` or `Shift` to initiate interactions in a desktop environment. Such key actions are not available on mobile devices, so if your custom widget supports such key actions on a desktop, you must provide an alternative to function keys when using it on mobile devices. Alternative user interactions should be based on and follow the guidelines and standards for user interaction for mobile devices.

In Listing 5.35, you can see a custom widget that executes `doAction` after a user interaction. The specific user interaction depends on whether the custom widget is used on a mobile or nonmobile device. Such interactions are implemented as follows: For a custom widget running on a nonmobile device, you must click on the custom widget and press `Shift` at the same time to trigger the operation. For a custom widget running on a mobile device, you need to tap the custom widget with two fingers to trigger the operation.

```
class ColoredBox extends HTMLElement {

    constructor() {
        ...
        this._myEventListenersRegistered = false;
    }

    onCustomWidgetBeforeUpdate(changedProps) {
        this._props = { ...this._props, ...changedProps };

        // Registers the event listeners when not in design mode,
        // when the event listeners haven't been registered yet,
        // and when in mobile mode.
        if (this._props["designMode"] === false && !this._
myEventListenersRegistered) {
            this._props["mobileMode"] && this._props["mobileMode"] ==
= true ?
            this.registerTwoFingerTapListener() :
            this.registerShiftClickListener();
            this._myEventListenersRegistered = true;
        }
    }

    registerTwoFingerTapListener() {
        this.addEventListener("touchstart", event => {
            event.preventDefault();
            if (event.targetTouches.length === 2) {
                // At view time, touching with two fingers performs
the action below.
                this.doAction();
            }
        });
    }

    registerShiftClickListener() {
        this.addEventListener("click", event => {
            const shiftKey = event.shiftKey;
            if (shiftKey) {
                // At view time, shift-clicking performs the action
below
                this.doAction();
```

```
            } else {
                this.dispatchEvent(new Event("onClick"));
            }
        });
    }

    doAction() {
        // Do something.
    }
}
```

Listing 5.35 Modified coloredbox_main.js for Mobile Device Optimization

The example registers one of two event listeners when the custom widget is in view mode. An event listener that calls the `doAction` operation is registered when the custom widget is executed and clicked on a nonmobile device and the `Shift` key is pressed at the same time. If the custom widget is called up on a mobile device, an event listener that executes the `doAction` operation is registered when the custom widget is tapped with two fingers.

5.6 Exporting Custom Widgets

Exporting custom widgets

A custom widget behaves like other widgets in view mode. Just as your users can create a PDF of your story and export it, they can do it with your custom widget.

This may cause problems with the display of your custom widget during PDF export. Therefore, please do the following:

- Pay attention to the export restrictions and test them in advance.
- Set the `supportExport` property in the JSON file of your custom widget to `true`.

[«]

Supported JavaScript Libraries

When creating the PDF files, you use JavaScript PDF generators and JavaScript libraries, which may not support all JavaScript libraries of your custom widgets. For example, diagram axes may not be exported when using *D3.js*, and there may also be obstacles with custom widgets with special CSS properties.

5.7 Best Practices

You can find helpful tips and tricks on various topics relating to custom widgets in the in the **Best Practices** section of the SAP Analytics Cloud Custom Widget Developer Guide (see *http://s-prs.co/v594408*). You'll also find the script files for the **Colored Box** custom widget there, and you can easily copy them if you want to create the widget yourself. In addition, SAP has published a variety of sample custom widgets on GitHub (see *http://s-prs.co/v594409*), and you can download them to help you understand step by step how more complex custom widgets are built.

5.8 Summary

In this chapter, we gave you a deep dive into custom widgets. We explained the concepts of custom widgets and how they are built, and we also covered how to host, upload, and delete a custom widget in SAP Analytics Cloud. We also walked you through a detailed example that illustrates how a custom widget operates.

In previous chapters, you learned everything about how to create stories and extend them with scripting capabilities, and in the next chapter, we'll take a look at how to operate, provision, and manage stories in SAP Analytics Cloud.

Chapter 6
Provision and Operation Stories with Scripting

This chapter looks at various options for providing stories in general, the enhancement of stories through scripting, and operational issues such as performance monitoring and notifications of updated data.

6

What you can expect

In Section 6.1, "Sharing Stories within SAP Analytics Cloud," we'll look with you at the various options for sharing stories with other users. You can share stories via the file system or via the SAP Analytics Cloud catalog, and there are additional setting options for the mobile app. In Section 6.2, "Embedding Stories in Websites, we'll explain how you can embed your story in other websites. Section 6.3, "Performance Monitoring," takes up a large part of this chapter and is dedicated to performance monitoring. Since scripting is much more complex and specialized than simple stories, performance problems are more likely to occur. Finally, in Section 6.4, "Best Practices for Performance," we provide some onboard tools and options for detecting and limiting potential performance problems.

6.1 Sharing Stories within SAP Analytics Cloud

Sharing stories

First, we'll take a closer look at sharing and managing stories in SAP Analytics Cloud. In particular, we'll look at the various management and provisioning options via the catalog or the file system in SAP Analytics Cloud, and then, we'll take a look at lifecycle management. A multilevel system landscape is particularly important for scripting, which means that content must also be transported between these systems.

6.1.1 Sharing and Managing Content

There are basically two ways to share stories: via the file system and via the analytics catalog. There's also a third option, in which you don't share the story itself but instead send the content of your story as a PDF file. We'll now take a closer look at these three options.

File system

File System

In the **Files** area of SAP Analytics Cloud, you have the option of sharing your story with other users (see Figure 6.1), who can be individuals or teams.

[»]

Prerequisite for Sharing Stories

Stories can only be shared with users if they have been granted the appropriate authorization in advance.

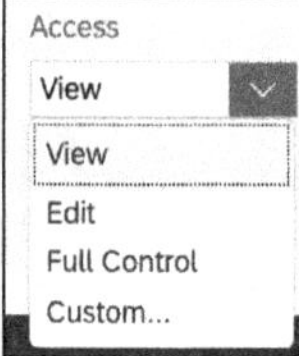

Figure 6.1 Suboptions for Story Access

There are four different authorization options:

- **View**
  With this authorization, a user can read and copy the story as well as view and add comments.
- **Edit**
  With this authorization, a user can update and edit the story.
- **Full Control**
  With this authorization, a user can delete and share the story and delete the comments in the story.
- **Custom...**
  With this authorization, a user can adjust the individual predefined authorizations (see Figure 6.2).

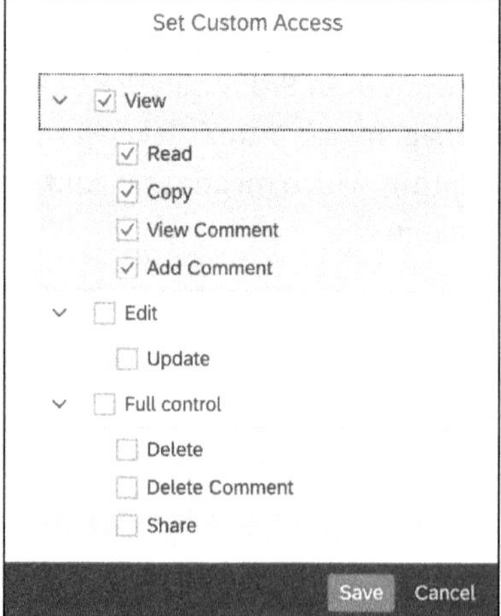

Figure 6.2 Setting Options for User-Specific Access

If you share a story in the file system, you also have the option of customizing the URL to be shared by using the **Customize link** button.

Creating your own URL

You can enter URL parameters and partially replace the URL with keywords so that it's shorter and therefore easier to remember. You can see an example in Figure 6.3.

The URL *https://demo.eu10.sapanalytics.cloud/sap/fpa/ui/tenants/9c5ff/bo/application/1AB01886262CB3AD71215EF548463BB0?mode=present* becomes *https://demo.eu10.sapanalytics.cloud/link/Demo*. The new URL is significantly shorter and therefore easier to remember.

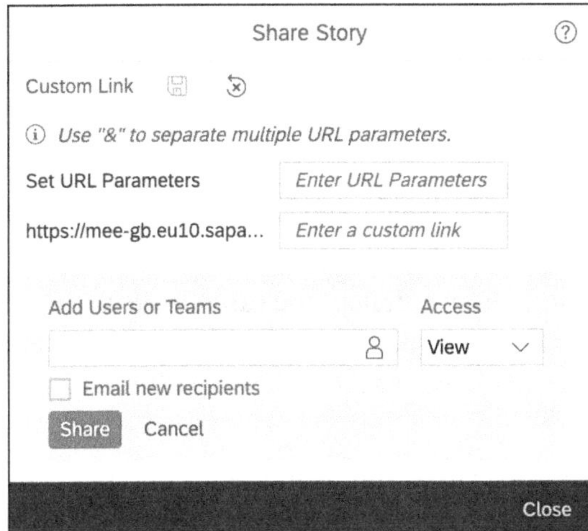

Figure 6.3 Customizing URL

You also have the option of informing the recipient by email about sharing the application or specifying a global bookmark as the default. As the application designer, you can also give your users the option of opening the dialog box for sharing the runtime:

```
Application.openShareApplicationDialog();
```

Analytics Catalog

Share stories with the analytics catalog

You can also share your story via the *analytics catalog* of SAP Analytics Cloud. This can be activated by your administrator, and if it's activated, you'll find a corresponding **Catalog** tab on your start page (see Figure 6.4).

The catalog is the central access point within SAP Analytics Cloud. You can customize the individual catalog tiles, add tabs, and define your own filter criteria.

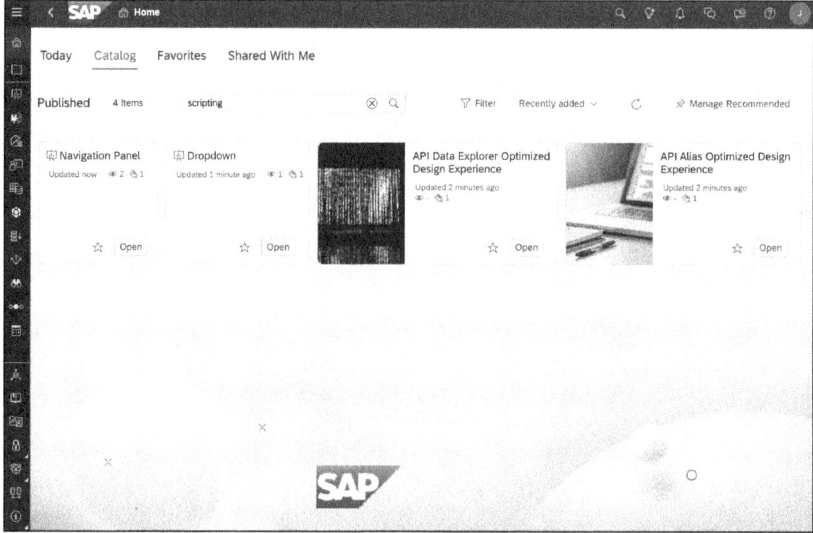

Figure 6.4 Overview of Catalog

If you want to publish your story in the catalog, you can release it at team level but not at user level. You can also specify whether you want to grant or deny read access (see Figure 6.5), which has the following effects:

- **Granting read access**
  If you grant read access, teams with it can open the tile and the story behind it.
- **Denying read access**
  If you deny read access, users from the team will see the tile in your catalog but won't be able to access the story. However, they can send access requests to you via this tile.

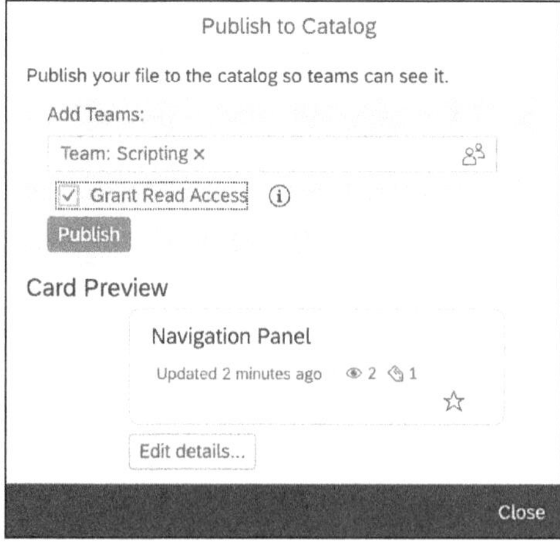

Figure 6.5 Publishing Story in Catalog

[«]

All Authorizations Must Be Assigned Correctly

In addition to read permission in the catalog, users need permission to read the story and the associated models.

Publication of Stories

Sending story

6

Another way to share your story is to publish it. This involves converting your story into a PDF document and sending it by email to SAP Analytics Cloud users and external persons.

The frequency of the possible dispatch times depends on your licensed number of SAP Analytics Cloud users. You can check the current number of available releases in your system overview (see Figure 6.6).

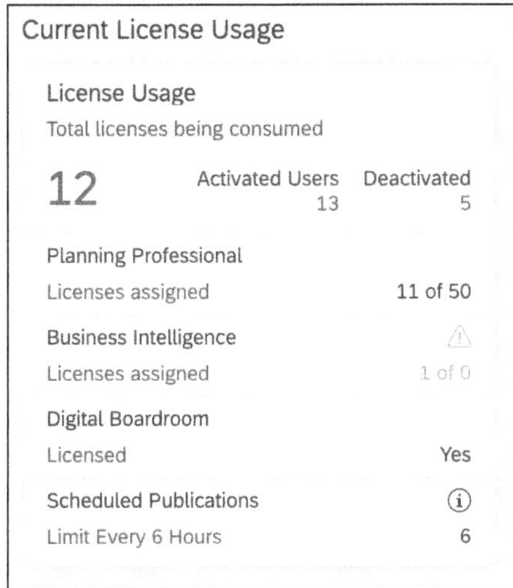

Figure 6.6 Overview of Available Publications

You can schedule this publication at a single predefined time or repeatedly at regular intervals, and you can also adjust available script variables.

[«]

Limited Dispatch to External Persons

You can send publications to external persons at no more than three email addresses.

To be able to use this function in your story, you must create a technical object for PDF export (see Figure 6.7). In this object, you can select the included widgets and make the following general settings:

- **Name field**
  You can use this field to assign a name to the exported file.
- **Paper Size field**
  This field allows you to choose between predefined sizes, or you can use the **Automatic** option.
- **Orientation field**
  This field allows you to choose between landscape and portrait format.
- **Page Details field**
  You can use this field to select whether page numbers should be displayed and, if so, whether they should be displayed in the header or footer.
- **Enable Export in the Background indicator**
  You can use this indicator to decide whether the export should take place in the background. This may result in longer waiting times.
- **Fully Display the Exported Widgets indicator**
  You can use this indicator to activate printing as a report. Only the table is currently supported. You can also decide whether you want to activate the **Header**, **Footer**, Appendix, and **Comments** elements.

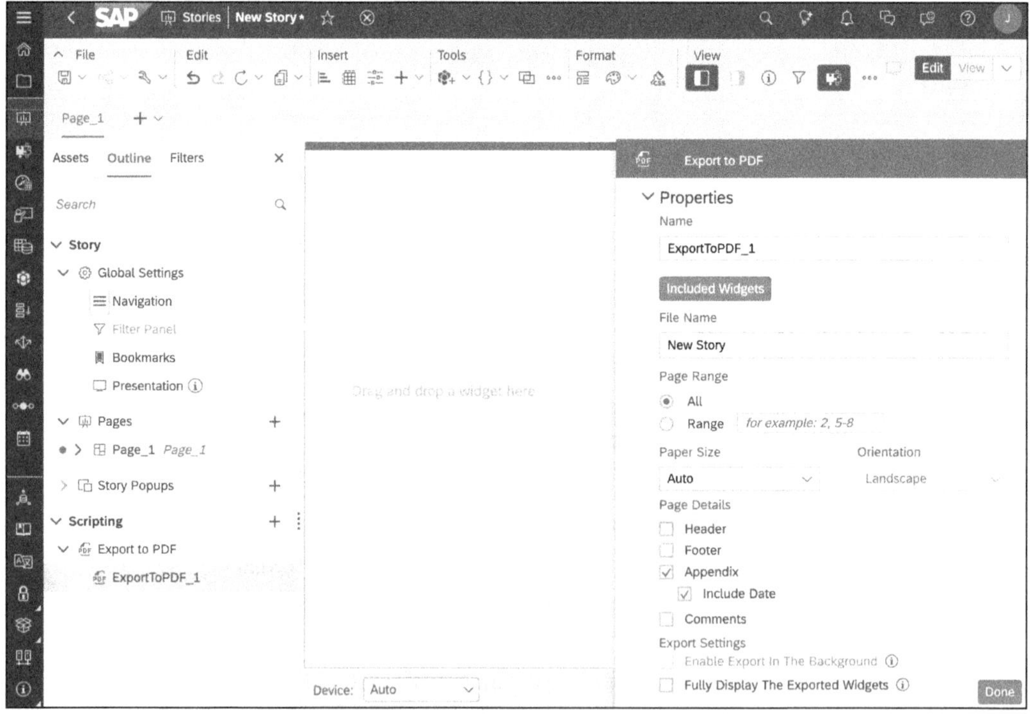

Figure 6.7 Exporting Object as PDF File

To send a story via a publication, you have the following two setting options, which can be found in the **Format** section of your graphics area (see Figure 6.8):

- **Automatically generate PDF export**
  This option is the default setting, in which the PDF document is generated automatically as soon as the rendering of the story has been completed. A story is fully rendered as soon as the last line of the `onInitialization` event has been executed. With this option, you don't need to write a script.
- **Manually generate PDF export via API**
  You can use this option if you want to make advanced settings with scripts and control the time of the export more precisely. Here, you can, for example, insert a script in your `onInitialization` event at the desired point:

```
if (Application.isRunBySchedulePublication()) {
     Scheduling.publish();
 }
```

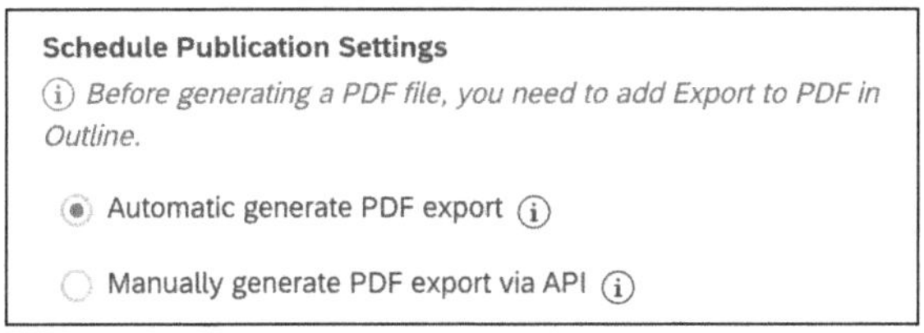

Schedule Publication Settings

ⓘ *Before generating a PDF file, you need to add Export to PDF in Outline.*

◉ Automatic generate PDF export ⓘ

○ Manually generate PDF export via API ⓘ

Figure 6.8 Setting Options for Publication

[«]

Mobile App

In addition to using the aforementioned three options, you can make your story available in a mobile app.

Translation

If you share your story with many different people, it's likely that you'll want to make it available in different languages. This option is available to you for your story.

To enable translation for a story, you must activate the **Enable translation** switch in the story details (see Figure 6.9).

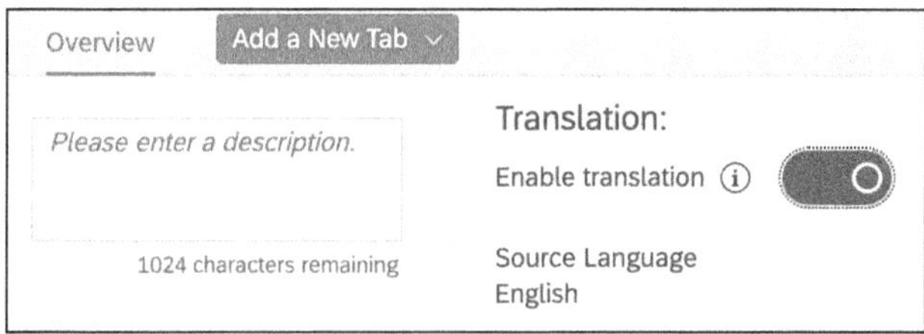

Figure 6.9 Activating Translation

As soon as you activate this option, you'll find your story in the **Translation** menu of SAP Analytics Cloud, and you can store the desired translations for the various elements (see Figure 6.10).

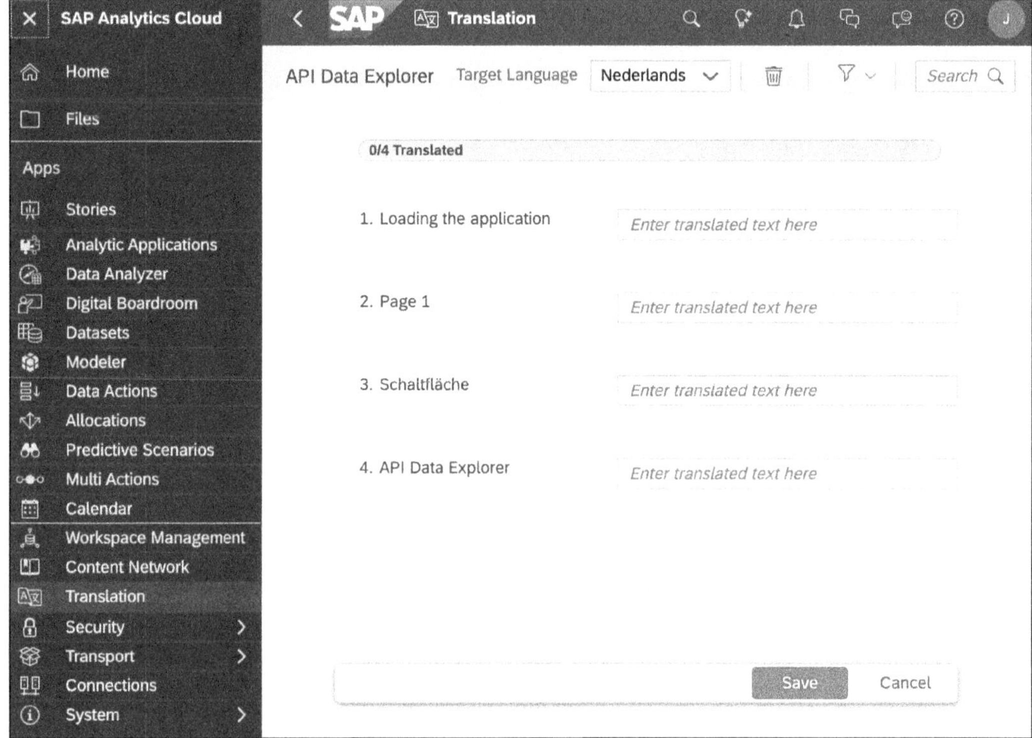

Figure 6.10 Translating Story

[»]

Scope of a Translation

Please note that only the elements of the story are translated; you must configure the translation of your data in your data source.

6.1.2 Lifecycle Management

Transports in SAP Analytics Cloud

Having many customization options increases the need for a multilevel system landscape in SAP Analytics Cloud. This ensures that changes to stories work correctly without affecting productive stories.

Suitable Release Versions

The target system must use either the same release version or a newer version of the source system. If the target system uses an older version, stories may not be supported.

There are two different ways of transporting objects between the different systems in SAP Analytics Cloud, and we'll look at them in more detail in the following sections. You can find both options in your SAP Analytics Cloud menu under **Transport**, where you have an area for **Export** and another area for **Import** (see Figure 6.11).

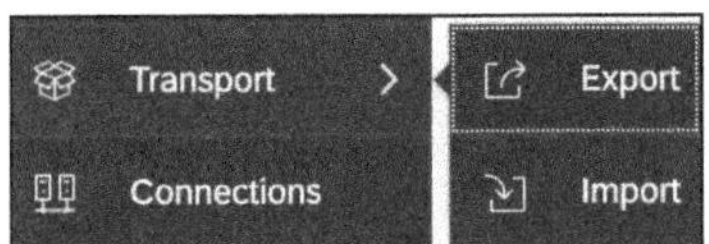

Figure 6.11 Menu Options for Transporting Objects

Transport via the Content Network

You can use the **Export** option to move your content via the cloud and share it with other systems.

Storage Space

A system can store up to 300 MB of exported content free of charge.

You must first create a new export. To do this, go to the **Transport • Export** path and create a new package there in the **Content network storage** area, using the **+** icon. You can now select which content should be included in the package (see Figure 6.12).

In the next step, you can enter a name, a description, and detailed information for your package. The destinations of this content are also defined here, where you can set which accesses each target should have (see Figure 6.13).

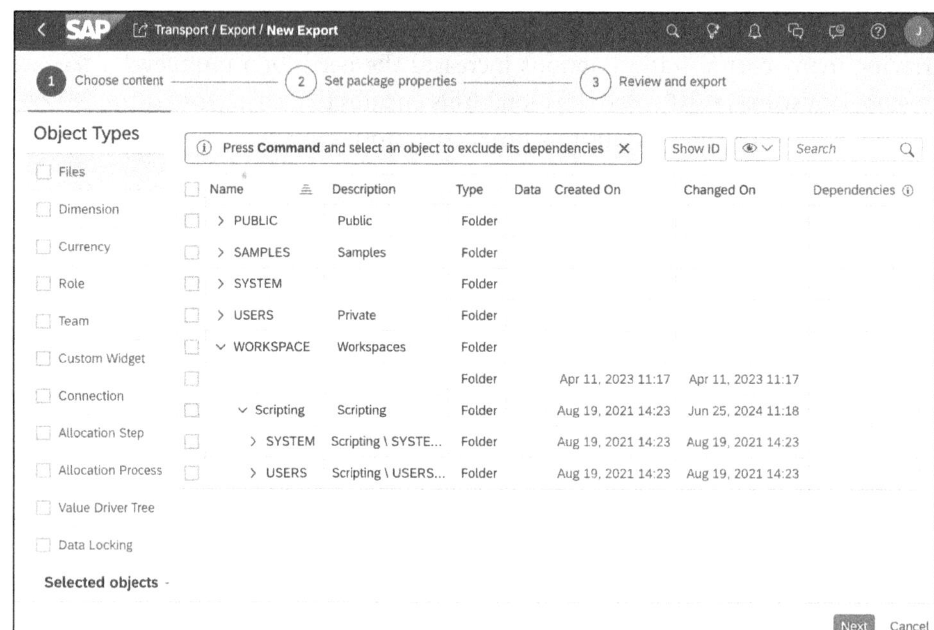

Figure 6.12 Selecting Content for Export

In the last step, you'll receive another overview of your content and settings so that you can check them. If everything is set correctly, you can start your export.

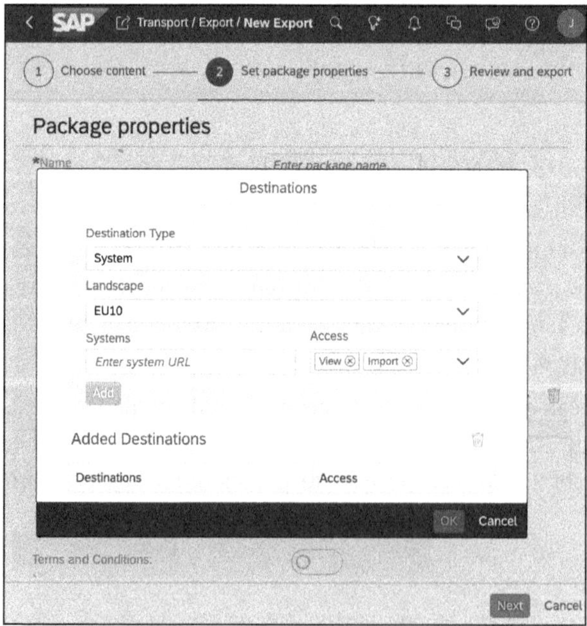

Figure 6.13 Setting Package Properties

You can track the progress of each export in your notifications. As soon as the export is complete, you'll receive a notification.

To import the content in this way, select **Import** under **Transport**. Here, you'll see all available content, as well as the name and a description of the content plus the URL of the tenant that provides the content (see Figure 6.14).

When you select content, you have the same options that you already know from importing from the public content network.

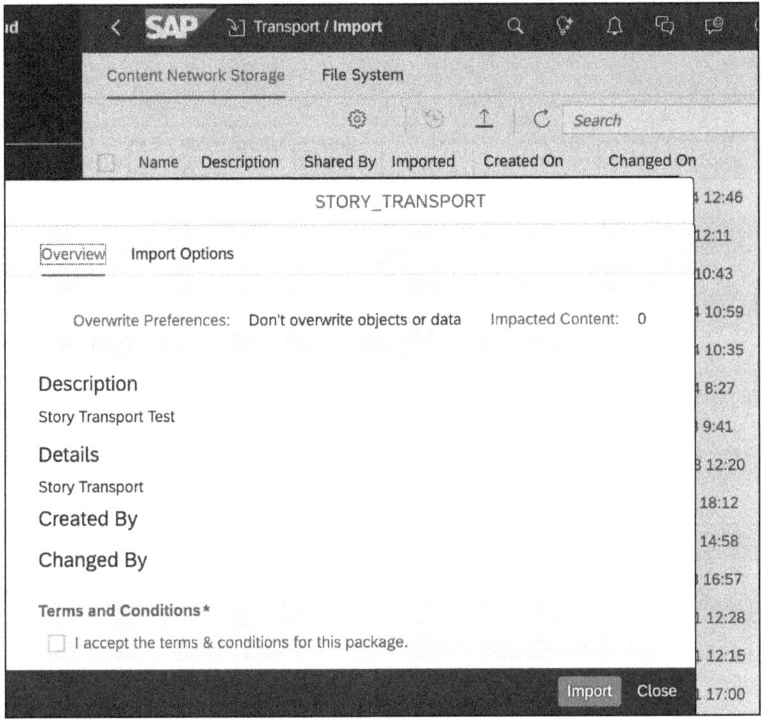

Figure 6.14 Importing Content via Private Content Network

Transport via the File System

Another option is transport via the file system, where the contents are moved between the systems via TGZ files. Here, you'll also see an overview of previous exports and information. Use the **+** icon to create a new export.

In the next step, you can select the desired content to be included in the export in an overview.

> **Selectable Contents**
>
> You can only access the public folder in the **Files** area.

The selected objects are now displayed in an overview on the right-hand side. You can start the export via the icon in the top right-hand corner, and you can then assign a name to your export (see Figure 6.15). The result is a TGZ file that you can import into your target system.

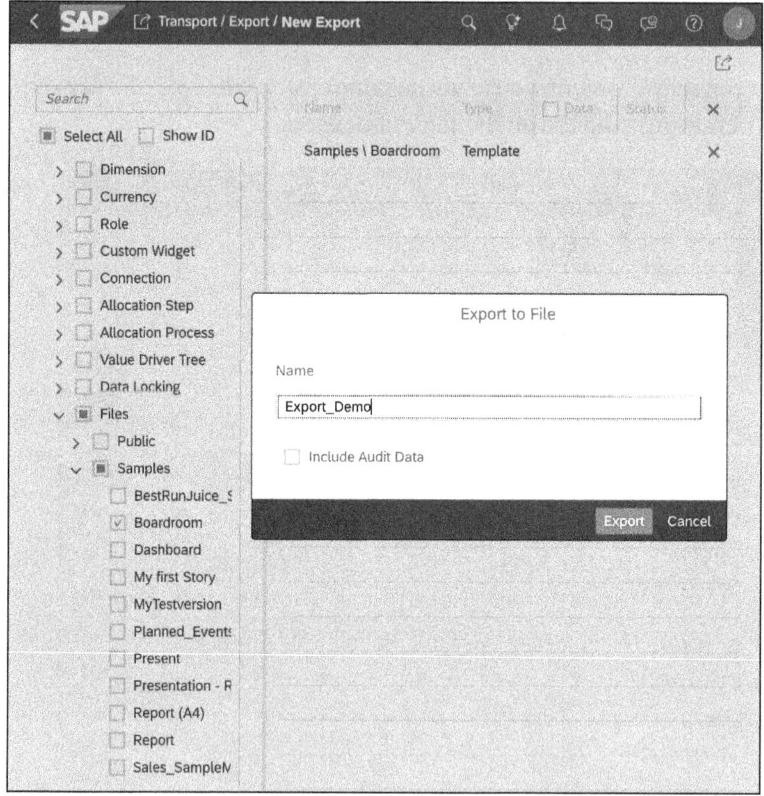

Figure 6.15 Export via File System

6.2 Embedding Stories in Websites

Integration with other websites

In SAP Analytics Cloud, you have the option of embedding your story in websites, and you can also embed websites in your story. To do this, you use the website widget, which you can use to establish message communication between the specific host and the embedded object. These two scenarios can be implemented using the APIs for sending messages, as we'll discuss in the following sections.

6.2.1 Embedding a Story

To embed a story in an HTML page, use an *iFrame* (see Figure 6.16), which is an HTML element that is used to structure websites. First, you must add the story as a trusted source address, and you can find the corresponding settings under **System Administration • App Integration • Trusted Origin Addresses**.

You can now trigger communication between the HTML page and the story. To do this, use the `postMessage` method:

```
Application.postMessage (receiver: PostMessageReceiver, message:
string, targetOrigin: string);
```

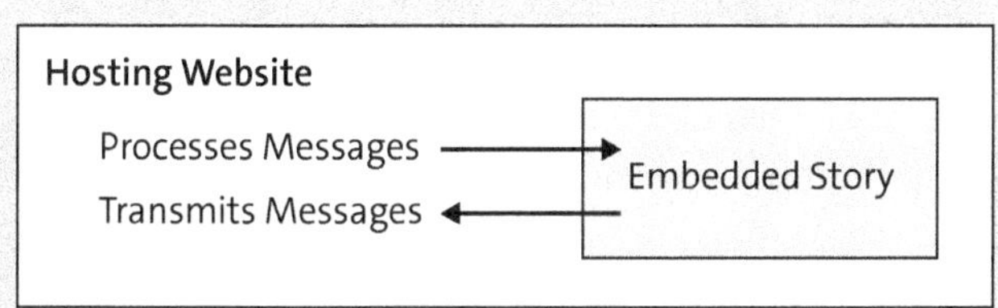

Figure 6.16 Embedding Application in Website

You can also use the `onPostMessageReceived` event to process messages that have been sent from the hosted website.

Checking the Address

Check the source address when receiving messages; you could be receiving data from a malicious website.

6.2.2 Embedding a Web Application

You can also embed web applications in your story and establish communication via the website widget (see Figure 6.17).

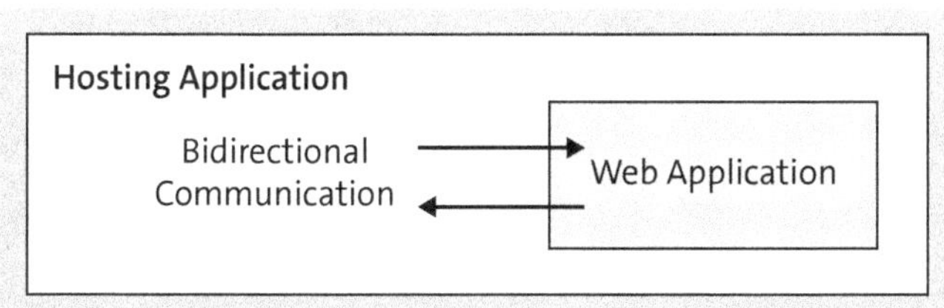

Figure 6.17 Embedding Website in Application

You can use the `postMessage` method to send messages from your story to the web application:

```
WebPage_1.postMessage(message: string, targetOrigin: string)
```

If you send the messages via the hosting story, their processing depends on the embedded application:

- In an SAP Analytics Cloud application, the receipt is processed via the `onPostMessageReceived` event.
- In a web application, you can process the message via the `window.on("message")` event.

However, if the hosting story wants to be received by the embedded website, processing takes place with the `onPostMessageReceived` event.

For an embedded story, messages can be sent using the `postMessage` method; for a web application, you must use the `window.parent.postMessage` event.

6.3 Performance Monitoring

Monitoring performance

There are many customization options in your dashboards related to creating stories with scripting in SAP Analytics Cloud. However, customization can lead to performance losses due to poorly written code or interactions between different lines of code. In this section, you'll learn how to find out where you lose your performance step by step and get to know some best practices.

SAP Analytics Cloud offers various tools to check the performance of your story and to uncover potential "brakes." Of course, you can also use the underlying models of this application to build your own stories or adapt existing applications, and we'll take a look at those in the following sections. Go to **System • Performance in** the navigation bar to open the tools (see Figure 6.18).

Figure 6.18 Performance Tools Menu

6.3.1 Performance Analysis Tool

Performance analysis

We start with the *performance analysis tool*, which can help you resolve performance problems. You can use the following decision path for orientation (see Figure 6.19).

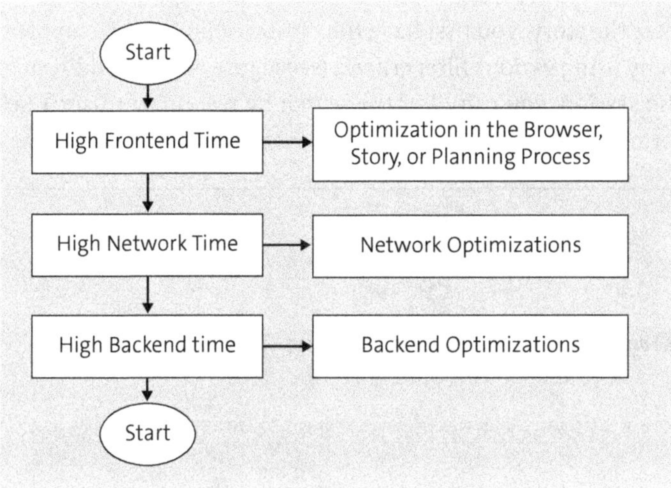

Figure 6.19 Decision Path for Optimization

Making optimizations

To find the point at which you can or should make optimizations, go through the decision path. There are three main categories in which there may be a need for optimization:

- Browser, story, application, or planning process
- Network
- Backend

If the front-end time is particularly high, you should first check your browser and client. Check whether the problem persists once you've closed all other tabs and applications. Are you using a virtual machine? If so, check whether the problem also exists on a physical computer. Your bandwidth or a third-party proxy can also have a negative impact.

Once you've made sure that the slow performance is not due to your browser or client, you should take a look at your story. If you've acted in accordance with best practices, or if, for example, a large number of widgets are already loaded when the story is initialized, then even though they are not needed here, you should optimize them and just load the needed widgets.

Network

If the network time is particularly high, you should check your reverse proxy, the VPN, and your network. If possible, check whether the problem is reproducible if you access the network without a VPN. Problems in your SAP data center can also lead to delays, so check the status in SAP Trust Center too.

Backend If your workload on the backend is particularly high, you won't have a problem arising in SAP Analytics Cloud, and you need to take a closer look at your data source. The best way to do this is to make several queries and try to narrow down the problematic query. Once you've found a critical query, look through it join by join and then make optimizations where necessary.

Application When you enter the story, you first have the option of limiting the upcoming results set by using various filter criteria (see Figure 6.20). In addition to the date of the session, you can filter the search by resource or user. Click the **Search** button to display the sessions that match your search criteria.

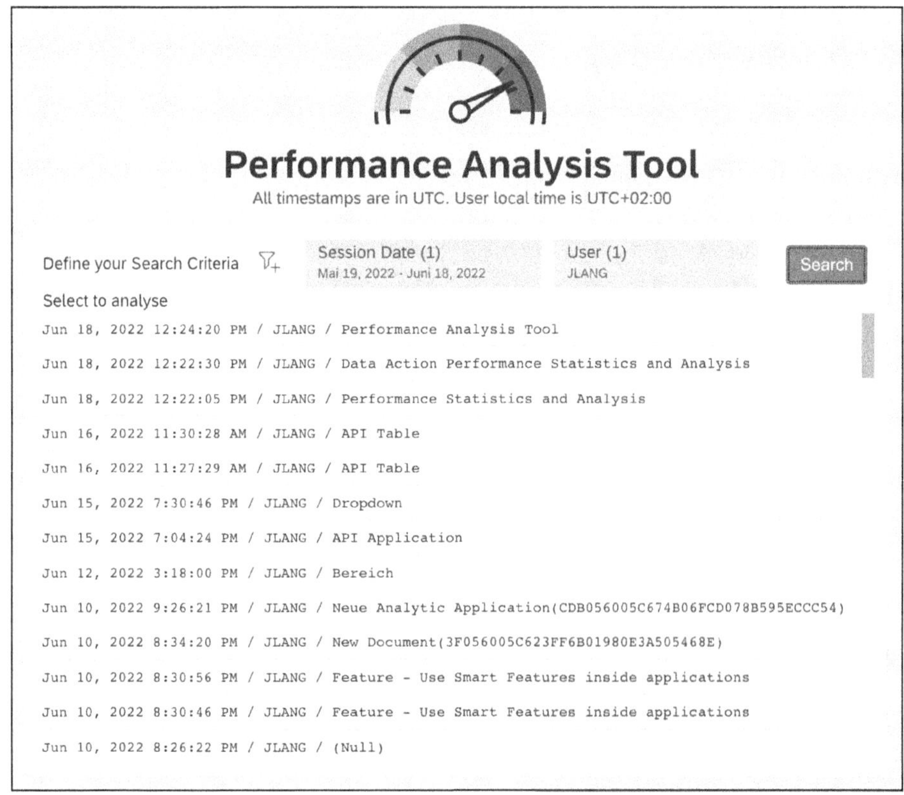

Figure 6.20 Getting Started with Performance Analysis Tool

You can now select the desired session from this results list.

In the view that appears, you'll see detailed information about this session (see Figure 6.21). You can then use this information to identify weaknesses or problems in your story. Among other things, the start time of the story, the five most time-consuming charts, and the models with the longest backend time will be displayed.

The lower section of the overview provides you with detailed information on the individual areas. You can take a closer look at the page load times, widgets, and general runtime. You can also filter these areas further (e.g., for a specific action or page, for a specific widget).

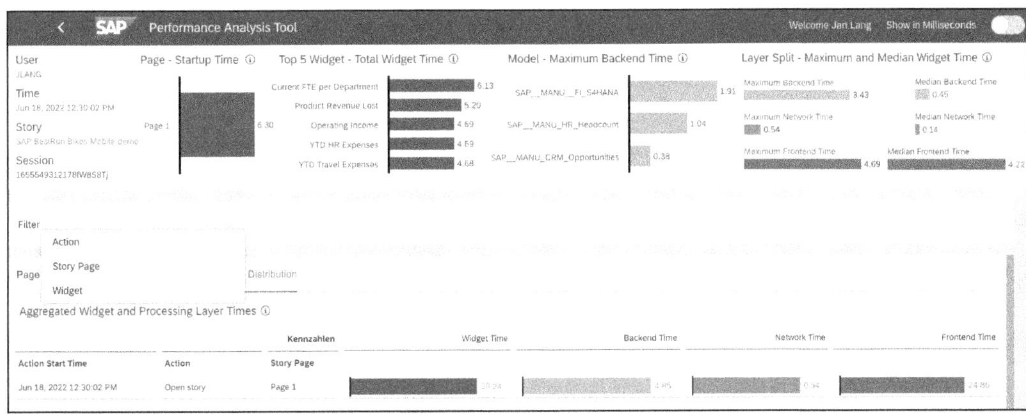

Figure 6.21 Detailed View of Performance Analysis

6.3.2 Benchmark Tool

Benchmark comparisons

The *benchmark tool* helps you determine whether your client or network is responsible for your performance problems (see Figure 6.22). To make this determination, you can run a client test or a network test.

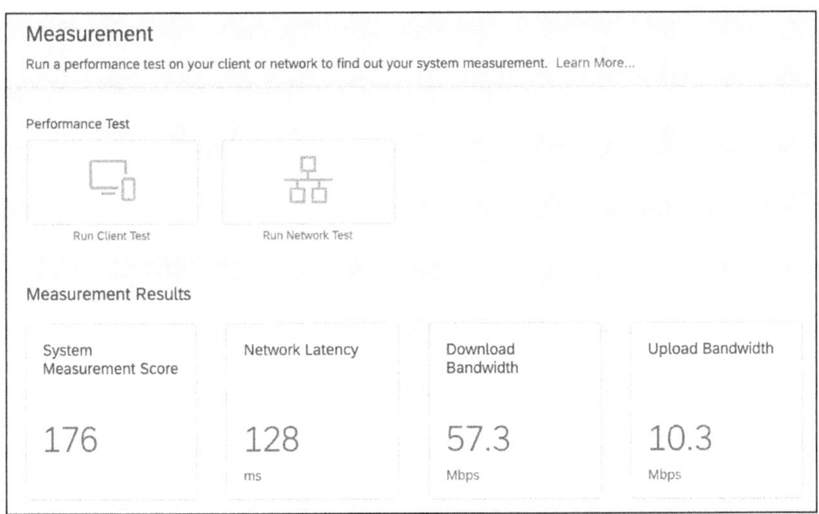

Figure 6.22 Benchmark Comparison

Benchmarking limits

The result of the *client test* is a *benchmark score*, which is calculated on the basis of several benchmark measurements. Benchmark scores fall into three categories:

- More than or equal to 75: Excellent
- Less than 75 but more than or equal to 50: Good
- Less than 50: Bad

To improve your benchmark score, you should pay attention to the CPU-intensive applications and processes on your computer that have a negative impact on the performance of SAP Analytics Cloud.

[»]

Optimal Conditions

Don't switch between browser tabs or minimize them during the test.

After the client test, you can run the network test, which returns three key figures:

- **Network latency**
  This key figure is the result of a calculation based on your latency and system bandwidth. The *latency* corresponds to the time that elapses between the request and the response from the SAP Analytics Cloud server.
- **Download bandwidth**
  This key figure indicates the amount of data transferred from the SAP Analytics Cloud server to the client in megabits per second.
- **Upload bandwidth**
  This key figure corresponds to the amount of data transferred from the client to the SAP Analytics Cloud server in megabits per second.

[»]

Activities during the Test

Bear in mind that several hundred megabytes can be downloaded during the performance test. Other download activities in the background can also have a negative impact on the analysis.

6.3.3 Statistics and Analysis

Statistics and analysis

This overview not only helps you to answer questions about your backend, but it also sheds light on some frontend topics. In the **SAP Analytics Cloud Performance KPIs** section, you'll find information on general KPIs for your system, such as active users, models, and the use of resources (see Figure 6.23).

The **Backend KPIs** section provides an overview of the backend KPIs. Here, you'll find information on average runtimes or the number of queries. The **Frontend KPIs** area is very focused on the number of logins.

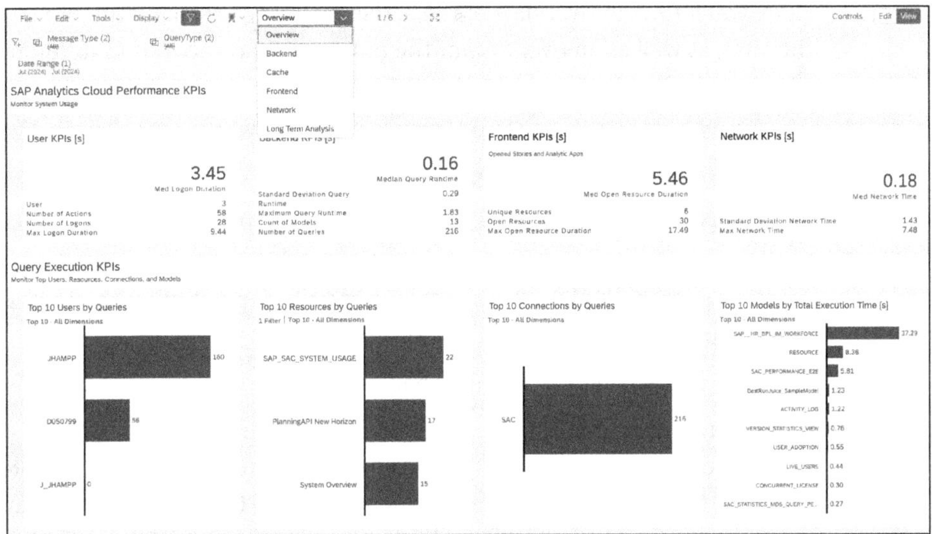

Figure 6.23 Overview of Performance Statistics

You can use a dropdown list to jump to other pages where you can find more detailed information.

6.3.4 Data Action Statistics and Analysis

Statistics on data actions

This story is very similar to the previously highlighted statistics and analyses, but the focus here is on planning and the data actions it contains (see Figure 6.24).

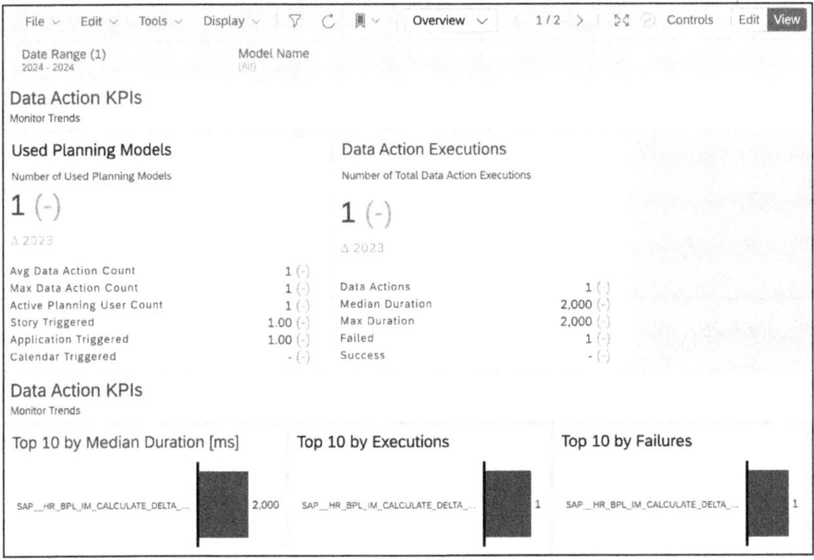

Figure 6.24 Performance Overview of Data Actions

Here, you'll find an overview of the active planning users and the models they use, as well as an overview of the number of data actions in different contexts, such as in the calendar or stories.

6.3.5 Popup to Analyze the Script Performance

Checking script performance

Another way to check the performance of your applications is via the *script performance popup*, which you can use to check the performance of your scripts at runtime. To be able to use this popup, you must run your story in an appropriate mode (see Figure 6.25).

There are two ways to access this mode:

- **By customizing the URL**
  To call up the story via the URL in this mode, you must add the `APP_PERFORMANCE_LOGGING=true` parameter to your application URL. Place the parameter before the # in your URL and precede the parameter with a question mark, as shown here:`/app.html?APP_PERFORMANCE_LOGGING=true#/`.
- **With the Script Performance icon**
  The **Script Performance** icon performs the manual steps just described for you. You can access this button via the change mode of your story, which you can find in the **Tools** area.

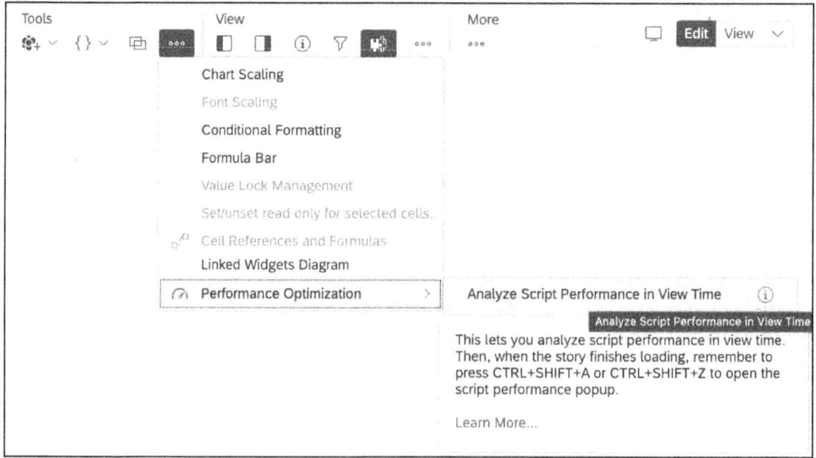

Figure 6.25 Calling Up Performance Popup

If you have entered the parameter in the URL or called up the story using the button and the application has finished loading, you can call up the popup using the key combination Ctrl+Shift+A or Ctrl+Shift+Z.

The popup with the loading times for the various scripts will then open (see Figure 6.26). If you move the mouse cursor over the bar, the exact execution time will be displayed.

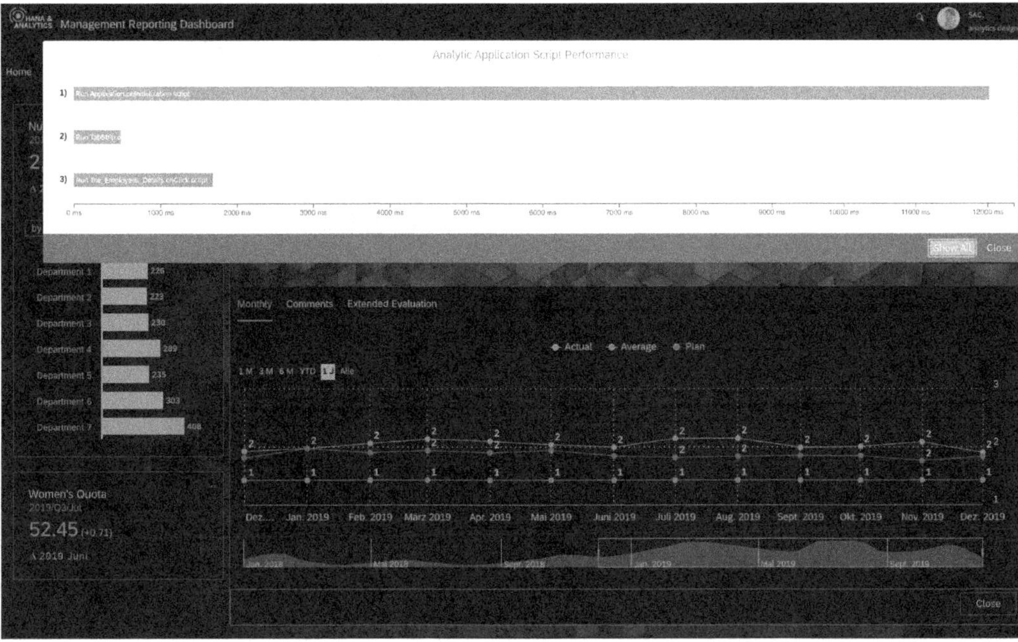

Figure 6.26 Displaying Performance Popup

If there are performance-relevant events, they are displayed in a red bar below the triggering event. There are currently two cases that can be displayed:

- Waiting for another widget in the background
- Calling up the description for an element of a dimension

6.3.6 Browser Development Tools and Debuggers

Use of the browser developer tools

You have the option of searching for and displaying your scripts in the development tools of your web browser, and we'll now show you how to do this, using the Google Chrome browser as an example. For a script to be found, it must have been executed at least once in the current session. You can open the Google Chrome development tools either by pressing F12 or via the Google Chrome menu. There, you'll find the development tools under **More Tools • Developer Tools** (see Figure 6.27).

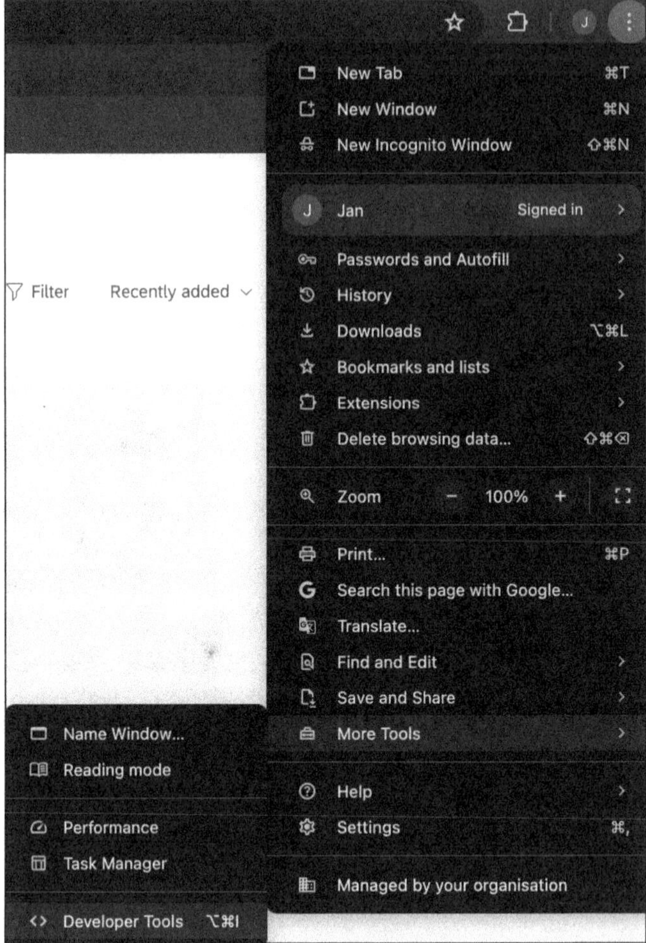

Figure 6.27 Opening Developer Tools

You are now in the developer tools of Google Chrome, where you can search for your scripts, either via a search field or via the tree structure (see Figure 6.28). To open the search field, press Ctrl+P on your keyboard, and you can then search for the name of your widget or function. Name suggestions are displayed during the search.

You can also search for your script using the tree structure on the left-hand side (see Figure 6.29). You can find it under **sandbox.worker.main**, and below it, you'll see your story and its elements.

You can also use the *debug mode* in the development tools of SAP Analytics Cloud. You start debug mode by adding the `debug=true` URL parameter to your story, and the script will be automatically stopped as soon as you reach a defined point. To include a breakpoint in your story, add the `debugger;` instruction to your code.

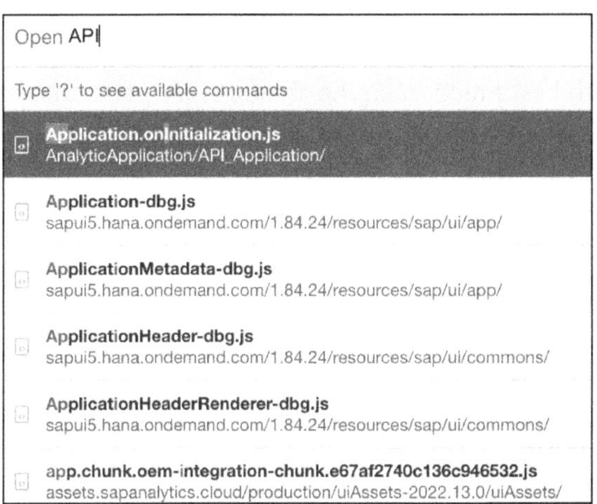

Figure 6.28 Search Bar in Developer Tools

Elements Console Sources Network Performance Memory Application Security Lighthouse Recorder Performance insights

Page Filesystem Overrides » | Application.onI...tialization.js ×

- top
  - mee-gb.eu10.sapanalytics.cloud
  - assets.sapanalytics.cloud
  - fonts.googleapis.com
  - sapui5.hana.ondemand.com
  - sandbox-iframe.html
- sandbox.worker.main.46798565bf715f586654.js
  - AnalyticApplication
    - API_Application
      - Application.onInitialization.js
  - assets.sapanalytics.cloud
- serviceWorkerProxy.js
- workerProxy.js

```
(async function anonymous(
) {
'use strict';
await this.$$.Application.showMessage(await this.$$.ApplicationMessageType.Success, "TEST");
console.log(await this.$$.Application.getInnerHeight());
await this.$$.Application.moveWidget(this.$$.Chart_1);
await this.$$.ApplicationMessageType.Success;
await this.$$.ApplicationMode.Present;
Array.isArray(true);
await this.$$.ArrayUtils.create(await this.$$.Type.boolean);
console.log(await this.$$.Application.getUserInfo());
console.log(await (await this.$$.Chart_1.getLayout()).getHeight());
await this.$$.LayoutValue.create(90, await this.$$.LayoutUnit.Percent);

//# sourceURL=file://AnalyticApplication/API_Application/Application.onInitialization.js
})
```

Figure 6.29 Menu Structure

You can find another option for defining a breakpoint in the developer tools. To do this, open the script to be stopped and select the line in the developer tools where it should stop. To remove a breakpoint, click on it again (see Figure 6.30).

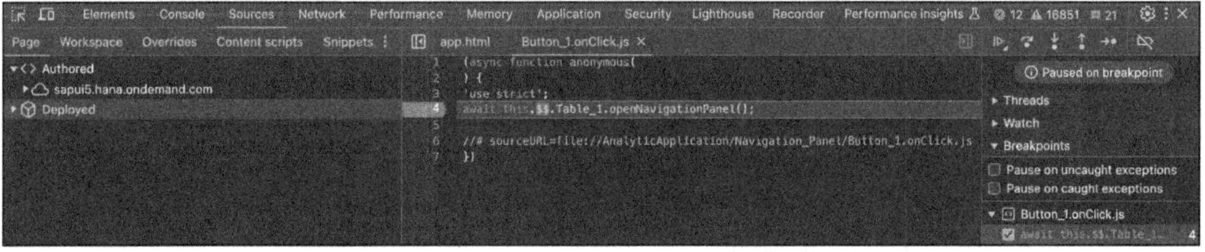

Figure 6.30 Stopping at Breakpoint

6.4 Best Practices for Performance

Recommendation

Now, we provide you with best practices in the area of story performance that you should consider when creating stories. In principle, all best practices from the simple SAP Analytics Cloud story are relevant.

6.4.1 Pausing Updates

When creating a story, you have the option of making settings for data updates (see Figure 6.31).

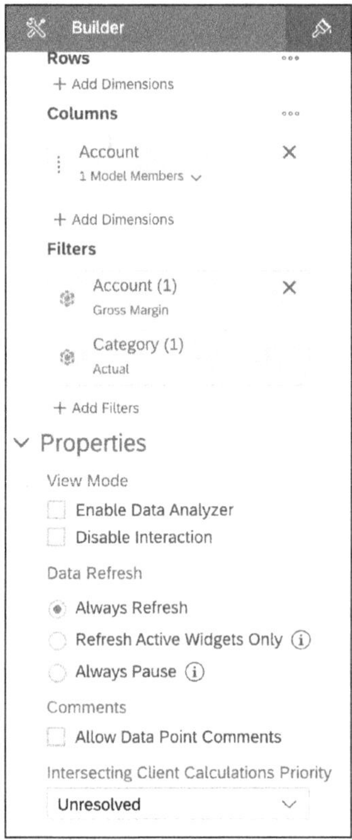

Figure 6.31 Data Updates in Builder

You can make these settings either in the builder or via an API. In this way, you can improve the performance of your story and avoid unnecessary data updates. Three different modes are available for this purpose:

- **Always Refresh**
  This is the default setting in every story, and in this mode, the chart is always updated.

- **Refresh Active Widgets Only**
  This mode offers the option of avoiding unnecessary data updates. Only visible widgets are updated, and the updating of invisible widgets is paused. As soon as an invisible widget becomes visible, all previously executed actions that lead to a data update are applied to this widget.
- **Always Pause**
  In this mode, no data is updated during runtime.

You can use the `RefreshPaused` API to give your users the option of starting and stopping data refreshes. The scripting overwrites the settings in the builder area.

[«]

Inactive Widgets

Widgets that are located in inactive tabs, book pages, and popups are treated as invisible.

6.4.2 Loading in the Background

Loading behavior of widgets

You can improve the performance of your story by setting widgets to load in the background. If this option is activated, it can improve the start performance of your story. As a result, all widgets visible at the start of the story are loaded. In the next step, the invisible widgets are initialized in the background.

The `loadInvisibleWidgets` URL parameter offers another option for calling up a story in the corresponding mode. It overwrites the setting in the story.

The `loadInvisibleWidgets` URL parameter can contain two different values:

- **`loadInvisibleWidgets=onInitialization`**
  This value forces the standard mode and loads all widgets at the beginning.
- **`loadInvisibleWidgets=inBackground`**
  This value activates the mode to load the invisible widgets in the background.

6.4.3 Widget Initialization at Startup

Initialization of widgets

With this option, you can decide whether each individual widget should be loaded directly when the story is initialized. This allows you to force the initialization of a widget at startup to use it in the `onInitialization` event without having to wait for the widget to initialize in the background (see Figure 6.32).

The widget is now initialized with all other visible widgets at startup, even if it's in the background. You can find the setting for this in the designer in the **Format** area.

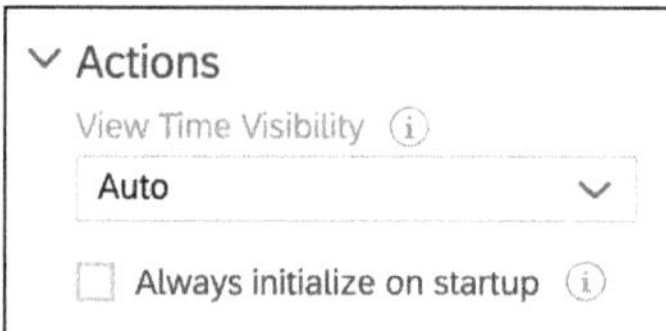

Figure 6.32 Forcing Initialization of Widget

6.5 Summary

In this chapter, you've learned different ways to provide your stories and how to interact with a web application or embed stories into a web application. The most important parts of this chapter were the sections where you learned how to detect performance pain points in your stories and scripts and follow best practices to avoid performance issues. In the next and final chapter, we'll give you our outlook on the SAP Analytics Cloud roadmap.

Chapter 7
Outlook

In this final chapter, we give you an outlook on the SAP Analytics Cloud roadmap.

7

In Section 7.1, "Roadmap," we'll introduce you to SAP's Roadmap Explorer, where you can find all up-to-date information about future features from every SAP solution. In Section 7.2, "Joule," we'll give you an overview of SAP's digital copilot and how it will be integrated into SAP Analytics Cloud.

7.1 Roadmap

Let's start with the *roadmap* and the expected innovations related to it. Before we begin, please bear in mind that the times at which these innovations are scheduled to appear can be postponed at any time, and the innovations can also be completely canceled.

At this point, we can only provide an outlook on the current status of the roadmap as of summer 2024. As the roadmap is subject to constant changes, the functional scope of individual roadmap items may change at any time.

Road Map Explorer

You can find an up-to-date roadmap in the *SAP Road Map Explorer* (see *https://roadmaps.sap.com/welcome*), which you can see in Figure 7.1. There, you can search for the desired product —in our example, you would enter "SAP Analytics Cloud" in the search field.

In the SAP Analytics Cloud overview, the cycles of the various planned releases appear in the individual columns (see Figure 7.2). Below this, you'll find various categories that are expected to be available in the specific release. You can expand these categories, and you'll then be shown an overview of the individual functions of the selected category in the form of tiles. You can then either expand these tiles again to obtain further information on the specific function, or you can click on the tile and a popup will open (see Figure 7.3). In this popup, you'll find detailed information on the desired function.

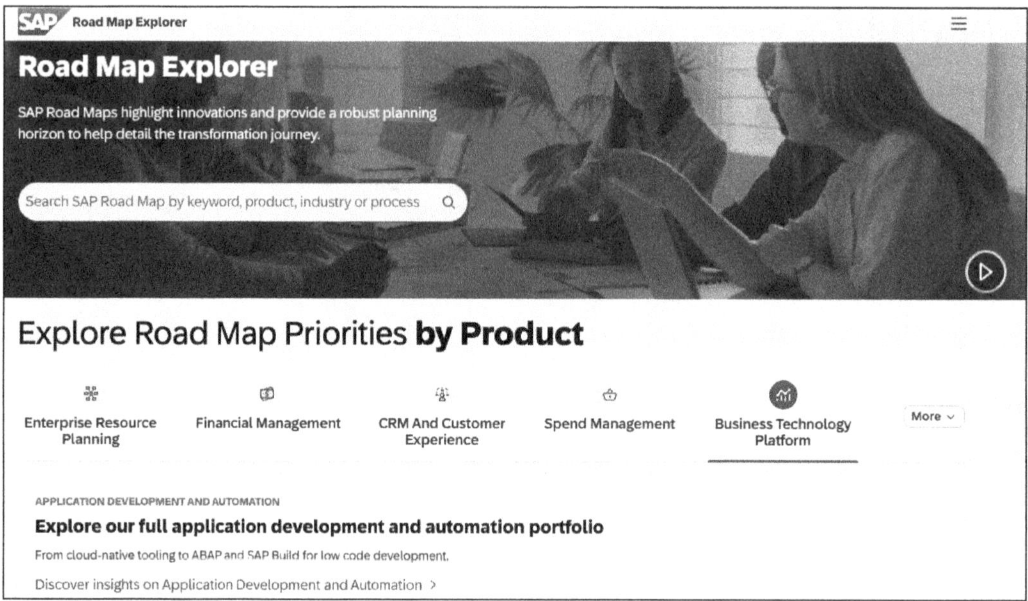

Figure 7.1 SAP Roadmap Explorer

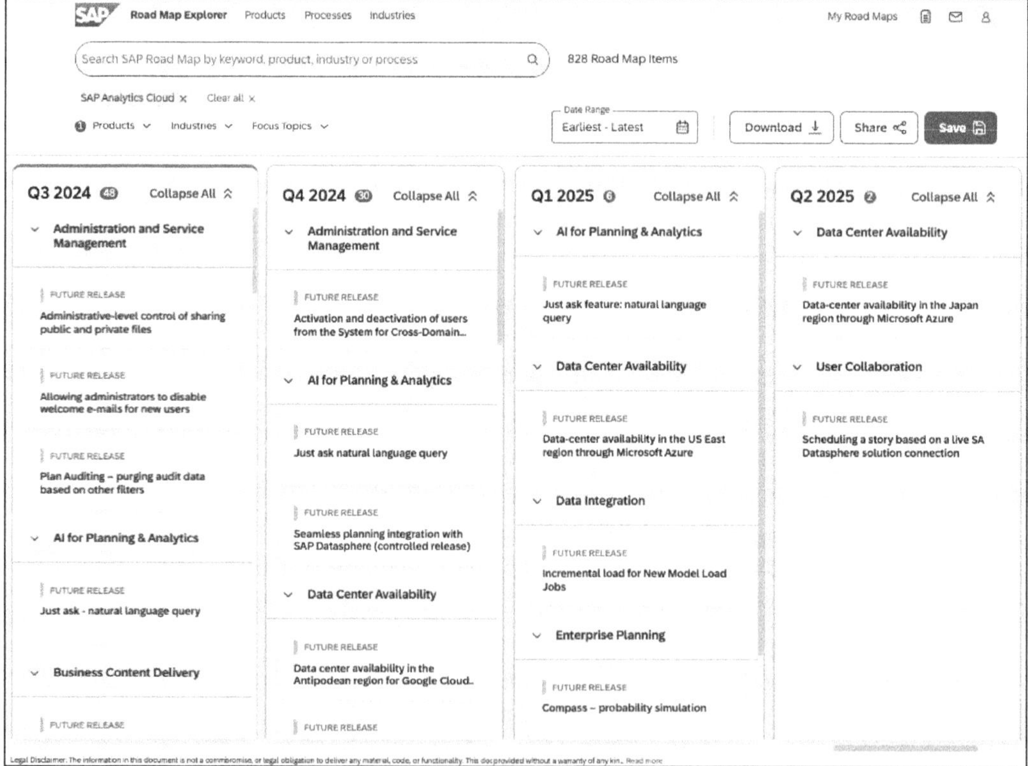

Figure 7.2 Overview of Roadmap for SAP Analytics Cloud

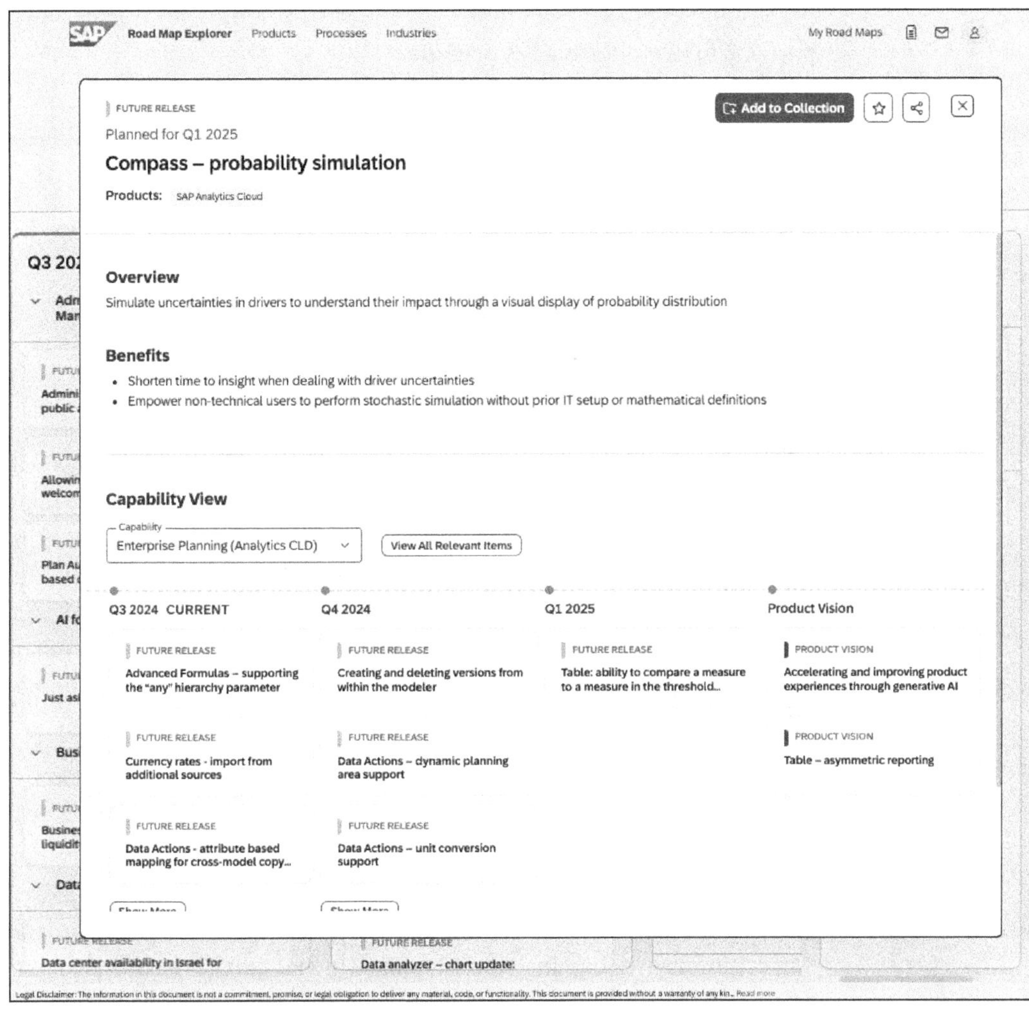

Figure 7.3 Detailed Information on Roadmap Item

We're sure that more functions will be added to those displayed in the Roadmap Explorer, as not all functions are large enough for their own roadmap item. The SAP Analytics Cloud solution will continue to grow in the future, and as in the past, many new functions will be added.

7.2 Joule

Joule

A major innovation for SAP Analytics Cloud will be the integration of *Joule* in the future. Embedding Joule in SAP Analytics Cloud is part of SAP's companywide strategy for Business AI. Joule is SAP's AI copilot that will revolutionize your interactions with SAP systems. For general information about Joule and the SAP Business AI strategy, please see *http://s-prs.co/v594410*.

In the context of SAP Analytics Cloud integration, Joule will be able to support you in the following key activities:

- Answering analytical questions faster
- Enriching data models with advanced calculations
- Discovering data and design data visualizations
- Autogenerating custom scripts to extend dashboards
- Simulating and analyzing driver impacts
- Creating and developing business plans
- Receiving best practices and how-to guidelines in context

You can also find this list at *http://s-prs.co/v594411* (subject to change).

Joule will also be able to translate natural language into scripts. This will speed up the creation of stories with scripting, and it will also provide open access to the scripting functionalities for story designers with little scripting experience. In the future, story designers will be able to interact with Joule to translate the logic of their algorithms into JavaScript. At SAP Sapphire 2024, SAP announced that Meta Llama 3 large language models will be used for this purpose (see SAP Sapphire 2024 News Guide at *http://s-prs.co/v594412*).

7.3 Summary

In this chapter, we first gave you an overview of SAP Roadmap Explorer, which is your single point of access to the latest roadmap information on SAP solutions. We then introduced Joule, which is part of the SAP Business AI strategy that will change the way you interact with SAP solutions. We also highlighted how Joule can support you when you're working with SAP Analytics Cloud.

Appendix A
Important Literature

Here, we would like to draw your attention once again to the important literature and sources for the topic of story design with SAP Analytics Cloud. We used the sources we've listed here when writing this book, and the web-based literature in particular offers great added value, as these are official SAP publications that directly incorporate changes that are introduced via new releases. These sources are therefore very up to date, they delve deeper into the subject matter, and they offer you complementary additions to the content discussed in this book:

- Abassin Sidiq, *SAP Analytics Cloud* (Quincy, MA: SAP PRESS, 2020). This is the practical handbook for SAP.
- GitHub, "SAP Analytics Cloud, Custom Widget Samples," *http://s-prs.co/v594413*.
- SAP, "Help Portal (Documentation): SAP Analytics Cloud," *http://s-prs.co/v594414*.
- SAP, "Help Portal (Documentation): SAP Analytics Cloud Custom Widget Developer Guide," *http://s-prs.co/v594415*.
- SAP, "Help Portal (Documentation): SAP Analytics Cloud Help, Scripting (Optimized Story Experience and Analytics Designer)", *http://s-prs.co/v594416*.
- SAP, "SAP Analytics Cloud, Analytics Designer API Reference Guide", *http://s-prs.co/v594417*.
- SAP, "SAP Analytics Cloud, Analytics Designer Developer Handbook", *http://s-prs.co/v594418*.

Appendix B
The Authors

Josef Hampp has been working on analytics and application development at SAP Deutschland SE & Co. KG since 2017. His focus is on the cloud offering of the SAP Business Technology Platform, and since his move to Customer Advisory, he has been working more intensively on the topics of data and analytics, including regulatory reporting and SAP's Business AI strategy. As a solution advisor, he is the point of contact for customers on these topics, and he also regularly represents SAP as an expert at public events such as conferences and webinars. Previously, he worked in the Innovation Services sales team, developing prototypes and dashboards with SAP Fiori and SAP Analytics Cloud. He studied business informatics with a focus on data science at Saarland University.

Jan Lang has been working with SAP Analytics Cloud since 2017. He started his career at FC Bayern Munich as the main contact person for the implementation of SAP Analytics Cloud. After interim positions as a team leader in the area of low-code development of SAP Fiori apps in the SAP and non-SAP environment at a consulting company and as a data and technology consultant (BI) at MHP, he returned to FC Bayern Munich in 2021. He is responsible for reporting with various SAP technologies in the IT & Tech department, and he supports customers of the FC Bayern Digital & Media Labs with SAP Analytics Cloud and SAP Datasphere.

Index